Geographies of Comfort

Bringing together conceptual and empirical research from leading thinkers, this book critically examines 'comfort' in everyday life in an era of continually occurring social, political and environmental changes.

Comfort and discomfort have assumed a central position in a range of works examining the relations between place and emotion, the senses, affect and materiality. This book argues that the emergence of this theme reflects how questions of comfort intersect humanistic, cultural-political and materialist registers of understanding the world. It highlights how geographies of comfort becomes a timely concern for Human Geography after its cultural, emotional and affective aspects. More specifically, comfort has become a vital theme for work on mobilities, home, environment and environmentalism, sociability in public space and the body. 'Comfort' is recognized as more than *just* a sensory experience through which we understand the world; its presence, absence and pursuit actively make and un-make the world. In light of this recognition, this book engages deeply with 'comfort' as both an analytic approach and an object of analysis.

This book offers international and interdisciplinary perspectives that deploy the lens of comfort to make sense of the textures of everyday life in a variety of geographical contexts. It will appeal to those working in human geography, anthropology, feminist theory, cultural studies and sociology.

Danny McNally is Lecturer in Geography at Teesside University. His research draws from cultural and social geography, and art theory and practice to explore pressing social and environmental issues. He has a PhD in Cultural Geography from Royal Holloway, University of London.

Laura Price is Research and Project Manager at PositiveNegatives based in SOAS, University of London. She is also co-editor of *Geographies of Making, Craft and Creativity* published by Routledge in 2018. Her research explores feminist geography, education, craft, and creativity.

Philip Crang is Professor of Cultural Geography at Royal Holloway University of London. He was editor of the journal Cultural Geographies from 1999 to 2008. His research is concerned with the material textures of places and the mobilities of people, things and ideas that constitute them.

Geographies of Comfort

Edited by
Danny McNally, Laura Price,
and Philip Crang

Routledge
Taylor & Francis Group
LONDON AND NEW YORK

First published 2021
by Routledge
2 Park Square, Milton Park, Abingdon, Oxon OX14 4RN

and by Routledge
52 Vanderbilt Avenue, New York, NY 10017

Routledge is an imprint of the Taylor & Francis Group, an informa business

British Library Cataloguing-in-Publication Data
A catalogue record for this book is available from the British Library

Library of Congress Cataloging-in-Publication Data
A catalog record has been requested for this book

ISBN: 978-1-4724-5402-7 (hbk)
ISBN: 978-1-315-55776-2 (ebk)

Typeset in Times New Roman
by codeMantra

Contents

Figures

Notes on contributors

Jamie Adcock has a PhD in Cultural Geography from Royal Holloway, University of London. His thesis explored the development of children's bedrooms from the late nineteenth century to the present day, and examines the bedroom's relationship with different genres of text. The thesis weaves together his broad research interests in children's geographies, domestic geographies, historical geographies, material culture and geographies of consumption.

David Bissell is Associate Professor and Australian Research Council Future Fellow in the School of Geography. He combines qualitative research on embodied practices with social theory to explore the social, political and ethical consequences of mobile lives. His research draws on cultural geography and mobilities research to investigate contemporary social problems involving mobility-labour relationships.

Maria das Graças Brightwell (1967–2018) was Postdoctoral Researcher at Universidade Federal de Santa Catarina. Her research is situated at the intersections of social, cultural and development geography and animated by interrogating processes of food production and consumption in a variety of settings. Her PhD in Human Geography was from Royal Holloway, University of London.

Eliza De Vet is Senior Research Assistant in the School of Geography and Sustainable Communities at the University of Wollongong. Her research has focused on individual and household experiences of everyday weather, climate (change) and natural hazard-related disasters, with a particular interest in mitigation and adaptation strategies. She has a PhD in Cultural Geography from the University of Wollongong.

Andrew Gorman-Murray is Professor of Geography at Western Sydney University. He is a social, cultural and political geographer. His research agenda is geographies of justice, encompassing the following areas: (1) gender, sexuality and space; (2) urban and regional transformations with respect to social and cultural diversity; (3) household dynamics with respect to home/work interchange and mobile work; (4) disaster planning and emergency management; (5) health, wellbeing and place; and (6) visual art and geography. Across this agenda, his interests include home and belonging; mobilities and place-making; emotional geographies and mental wellness; visual and material cultures; and social impacts of natural hazards and climate change. He is interested in qualitative, quantitative and visual approaches.

Mark Jayne is a Human Geographer working at the intersections of consumption, production, regulation, governance and policy. He is author of *Cities and Consumption* (Routledge, 2005) and co-author of *Alcohol, Drinking, Drunkenness: (Dis)Orderly Spaces (Ashgate, 2011)* and *Rethinking Childhood: Families and Alcohol* (Ashgate, 2013).

Weiqiang Lin is a Human Geographer at National University of Singapore. Employing cultural geographical approaches, his work examines the myriad ways in which contemporary society is mobilised through the creation of particular regimes and artefacts. His work opens up the cultural dimensions of mobilities production through an infrastructural lens. His recent research incudes the production of airspaces in Southeast Asia, discursive and technological framings of air logistics in Singapore and China as well as labour and automation in four of Asia's biggest international airports.

Jo Little is Professor of Gender and Geography at University of Exeter. Her main research and teaching interests are in rural communities and social identities and in gender and geography. She has been involved in research on women and the rural labour market for a number of years, looking at the constraints that impact on women's access to employment in the countryside and the relationship between the cultural construction of rurality and women's involvement in paid work.

Daryl Martin is Senior Lecturer in Sociology at the University of York. His research is interested in architecture, embodiment and health, new urban ruins, exurban sites on the edges of major cities and the role literature and literary theory plays in our understanding of the social and sense of place. He is co-director of the Centre for Urban Research (CURB) – an interdisciplinary gathering of academics, working across departments at York.

Stephanie Harel has a PhD in Social and Cultural Geography from Royal Holloway, University of London. Her research interests concern disasters, the geographies of home and emotion, post-disaster populations and return-migration. She is currently a Teacher of Geography at Wimbledon High School, London.

Katherine Morton is a Human Geographer researching the spaces, knowledges and practices associated with anti-ageing body work, considering the ways in which the ageing body is mediated through cosmetic technologies in beauty salons, spas and aesthetic clinics. Katherine's work is situated within feminist understandings of the body and contributes more broadly to geographical understandings of health and well-being.

Laura Prazeres is Honorary Associate Lecturer at St Andrew's University and has research interests in mobility, home, belonging, identity and qualitative methods. She was previously a Postdoctoral Researcher at the University of Dundee working on a DFID-funded project that explored youth transitions in protracted crisis. Prior to that, she worked as a Research Fellow on an ESRC-funded project on the dynamics of changes in international student mobility at the University of St Andrews. Laura has a PhD in Geography from Royal Holloway, University of London, which explored questions of home and identity within international student mobility.

Miranda Ward is a freelance writer, editor and lecturer based in the UK. She has a PhD in Cultural Geography from Royal Holloway, University of London, where her research centred on the geographies of lap swimming and the indoor swimming pool. She is particularly interested in the relationship between space, place and the body.

Daniel Webb has a PhD in Geopolitics from Royal Holloway, University of London, which explored the confluence of everyday geopolitical knowledges/practices and Christian theologies. Specifically, the thesis used ethnographic methods to explore the ways in which various Christian communities encountered, experienced and interacted with the geopolitical landscape in Jerusalem, Israel. He currently works as Higher Education Project Coordinator for the Refugee Support Network.

Acknowledgements

Laura, Danny and Phil would like to thank their generous and creative colleagues in the Geography Department at Royal Holloway, University of London (RHUL). This volume is indebted to an initial workshop and podcast recording on 'Comforting Geographies' at *Landscape Surgery* in 2013. It was shaped and strengthened by sessions at the Royal Geographical Society (with IBG) conference in 2013 sponsored by the Gender and Feminist Geographies Research Group and the Social and Cultural Geographies Research Group. The discussions, debate and feedback from these sessions shaped the collection and its agenda and we're grateful to feature papers within.

We would like to express our thanks to Oli Mould, Harriet Hawkins, David Gilbert, Sofie Narbed, Dan Webb and Jo Cagney specifically for their scholarship and support at RHUL.

The collection has been months and years in the making, in the end, as Laura and Danny navigated the precarity and discomfort of post-doctoral professional development. Throughout, colleagues at Ashgate and Routledge have been patient and understanding. We would like to thank Nonita Saha, Ruth Anderson and Faye Leerink for their support on the journey.

Danny would like to thank Avril Maddrell, Yasminah Beebeejaun, Katie McClymont, Brenda Mathijssen, David Bissell and Helen Wilson, who each gave encouragement, support and advice on this project. He would also like to thank his new colleagues at Teesside University, Sarah Perks, Ed Rollason, Elinor Morgan and Paul Stewart, who offered encouragement during the final stages of this book. Finally, he would like to thank his partner Annie, daughter Mollie and his family for the unrelenting love and support during this journey.

Laura would like to thank colleagues at the Royal Geographical Society (with IBG) and PositiveNegatives, SOAS for their understanding and support. In particular, Adarsh Muppane, Giulia Macgarr, Holly Sullivan, Matthew Briggs, Simon Pinfield, Claire Brown, Sara Wong, Emily Oliver, Benjamin Dix and Emma Parker. As ever, she is indebted to her family and friends for their continuous and unwavering support.

We dedicate this book to the memory of our colleague, Maria das Graças Brightwell.

1 Towards geographies of comfort

Laura Price, Danny McNally, and Philip Crang

Introduction

Historian John E. Crowley has documented that during the eighteenth century physical comfort was an innovative part of Anglo-American culture, defined as "self-conscious satisfaction with the relationship between one's body and its immediate physical environment" (1999: 750). The word "comfort" in this era became used increasingly to express this "satisfied" relationship, and indicated, according to Crowley, "a disposition to criticize traditional material culture and to improve it" (1999: 750). Centuries previous to this cultural-linguistic development, comfort mainly implied a "moral, emotional, spiritual, and political support in difficult circumstances" (1999: 751). Discomfort, at the same time, referred to feeling "sorrow, melancholy, and gloom rather than physical irritability" (1999: 751). Crowley goes on to explain that the consumer revolution in Anglo-American society during the eighteenth century developed a culture of comfort that integrated this new physical emphasis of the notion with the traditional one of moral support. Indeed, Brighenti and Pavoni (2019: 143) argue that the development of comfort-oriented societies is "the kernel of a theoretical genealogy of global capitalism". Fast forward to the consumer world of modern day, comfort is now also a quality attributed to material objects regarding whether they provide that sensory experience (Miller 2008, 2010; Vannini and Taggart 2014: 1078). For instance, scholars have explored the cultural economy of "the normalisation of comfort" in the home regarding heating practices (Shove 2003), and comfort as a negotiated sensibility through different types of clothing (Fuller and Bulkeley 2013; Johnson 2017).

The term "comfort" is omnipresent, both imbued with multiple and expansive meanings, whilst at the same time being everyday, ordinary and vernacular. This is not necessarily a book which seeks to define comfort – and its subsidiaries of discomfort, comfortable, comforting and so on. Rather, it is an exploration of the term as a geographical imperative. In this sense, while we do not present a concrete definition of comfort, we do wish to sketch out and illustrate the *geographies of* comfort. Cultural geographers have long attended the mundane, taken-for-granted aspects of society and life – seeking their ubiquity as worthy of study, investigation and engagement (Crouch 1989; Lorimer 2005; Laurier and Philo 2006; McNally 2015; Price 2015; Holmes and Hall 2020). Comfort, we argue, both develops and interrogates these agendas. As such, this volume compiles a selection of interventions that explore a range of geographies of comfort, bringing to the fore the textures and meanings behind them.

The *Geographies of Comfort* is the first volume to engage critically with "*comfort*" and "*discomfort*" as substantive concerns for Human Geography. It has gathered together conceptual and empirical research that deploys the lens of comfort to make sense of the textures of everyday life in a variety of geographical contexts. Comfort and discomfort have come to the fore in a range of works examining the relations between place and emotion, the senses, affect and materiality. We argue this emergence reflects how questions of comfort intersect humanistic, cultural-political and materialist registers of understanding the world. In this light, geographies of comfort are a timely concern for Human Geography after its cultural, emotional and affective "turns". More specifically, comfort has become a vital, emergent concern in a number of substantive fields within Human Geography, including work on mobilities (Bissell 2008; Adey *et al.* 2012); home (Brickell 2012a, b; Gorman-Murray 2012; Racz 2015); environment and environmentalism (Spinney *et al.* 2012; Hitchings *et al.* 2014; Vannini and Taggart 2013, 2014); sociability in public space (Eldridge and Roberts 2008; Boyer 2012); and the body (Colls 2012; Fuller and Bulkeley 2013; Johnson 2017). Geographers, anthropologists, sociologists and historians have recognized "comfort" as more than *just* an emotion through which we understand the world; rather, through its presence, absence and pursuit worlds are actively made and unmade. Advancing this recognition, this volume engages seriously with "comfort" as both an analytic approach and object of analysis.

This introductory chapter, split into four parts, sets out an argument for investigating the geographies of comfort. Section One, through the use of a series of illustrative examples, suggests that comfort and discomfort are everyday, common and ordinary ways of thinking about shared geographical experiences of world-making and un-making, mobility and socio-environmental transformation. As such, we believe they deserve critical reflection. In part, this means interrogating problematic and easy equations of comfort and discomfort with good and bad experiences or with conservative and progressive politics.

Section Two of the chapter contextualizes the geographies of comfort within wider disciplinary and interdisciplinary agenda. We do this by drawing out the different registers implicated in questions of comfort and discomfort. This includes the bodily, sensory and phenomenological geographies; cultural politics and cultural and feminist geographies; and affective materialism and non-representational geographies. In doing so, we seek to map out coordinates for a *fuller* understanding of the nuances, contradictions, distinctions and politics of "comfort" and "discomfort". In Section Three, we demonstrate how these different registers come together and are explored within the structure of the book across four substantive themes: bodies and environment; difference and encounters; materiality and textures; health and wellbeing. Finally, the chapter concludes drawing attention to the political importance and cultural geographies of comfort as world-making.

Negotiating slow (dis)comforting geographies of post-doctoral careers

> The neoliberal university requires high productivity in compressed time frames. Though the neoliberal transformation of the university is well documented, the isolating effects and embodied work conditions of such increasing demands are too rarely discussed.
>
> (Mountz *et al.* 2015: 1236)

The concept for this volume emerged out of a series of sessions on "Comforting Geographies" at the Royal Geographical Society with the Institute of British Geographers Annual Conference in 2013. There has been significant time between those conference sessions and the publishing of this volume, and we feel it important, particularly in the context of this topic, to highlight how our personal, everyday and on-going geographies of comfort have elongated the timeframe from conference to publication. In doing so, we seek not to be solipsistic but to make visible the difficulties, inequalities and systemic issues with the neoliberal university.

During the period of this book project two of the editorial team, Danny and Laura, went through the PhD process and into the post-PhD employment pool, moving between unemployment and temporary work for a number of years. Money was loaned, debt was accrued, and low-paid salaries were accepted out of necessity. Life was not *comfortable* in a material, social or emotional sense. As we sought to find time, energy and resources to complete the volume – we worried most about our responsibilities to co-authors, many were in a similar position. The Higher Education Statistics Agency (HESA) recorded that in 2018/2019 34 per cent of academics were employed on fixed-term contracts (HESA 2020). Our experience is symptomatic of the entrenched reliance on precarious labour within UK Higher Education, perpetuated through the sector's steady infiltration of neoliberal logics.

The *Geographies of Comfort* has been shaped and cultivated by the theories, thoughts and research of feminist cultural geographers, and we advocate here for "a feminist mode of slow scholarship works for deep reflexive thought, engaged research, joy in writing and working with concepts and ideas driven by our passions" (Mountz *et al.* 2015: 1253). Whilst getting "beyond your comfort zone" is critical and cerebrally integral to the generation of new ideas, thinking and doings (Eden 2014) – this, we argue, cannot be done sustainably without a degree of social, personal, emotional, material and personal comfort (security and familiarity) that can support the making of space in which research and career aspirations can flourish. It cannot be overlooked that systemically, the neoliberal university with its "regimes of harried time" does not facilitate for the feminist slow scholarship – indeed, it entrenches further the inequalities of geographies of comfort across race, gender, class, age.

Comfort *is* political, and we are grateful to have had, in the end, completed this volume. Our hope hereafter is twofold. First, that our co-authors' ethnographically and theoretically rich work inspires and advances the geographies of comfort agenda. Second, in illustrating the on-going politics of comfort as negotiation of everyday, personal and care*ful* geographies, it supports the slow scholarship challenge to

> neoliberalism's metrics and efficiencies, and instead recalibrates and changes academic culture. Our call is about more than simply making time for ourselves and our own scholarship; it is about collective action – big and small – in which we attend to the interpersonal and collective conditions that underpin knowledge production conducted with care.
>
> (Mountz *et al.* 2015: 1254)

(Dis)comfort and everyday life

The longest bench in the UK measures 324 metres long and sits 300 people. It is in Littlehampton, West Sussex. It is made from reclaimed hardwood, salvaged from landfill and old seaside groynes, colourful stacks, linked together, that twist and curve around lamp posts and street furniture, down coastal paths and crossings. It is hoped the bench will continually extend – more stacks will be added. You can pay, of course, to have your name engraved on one of those stacks – a historically meaningful act that often celebrates the ordinary enjoyment of people who have visited, rested and enjoyed a local area from the viewpoint of a particular bench. Rishbeth and Rogaly (2018: 1) argue that park benches, and architectural design more broadly, are intrinsic to understanding the micro-geographies of conviviality and care. The details of such design, they suggest, affect *both* the social and physical comfort and discomfort of people in place.

In this vein then, and rather contra in anecdote, let's picture another scene: spikes on a park bench. Bolts installed on shop doorsteps and windowsills. Even barbs on tree branches. These are examples of *hostile architecture* – urban design aimed at preventing people from accessing public space in ways deemed undesirable. So-called "anti-homeless spikes" intervene in the potential of public, civic space as a potential source of comfort for homeless people to linger, rest and sleep. It negates comfort through design in a way that is felt through the body in terms of aesthetic and tactility. These nefarious design decisions alter public spaces around the world, whilst symbolically and materially excluding vulnerable populations from particular places, often, based on ensuring the social comfort of *others* deemed undesirable. When we introduce or identify comfort there is *always* discomfort – it is a sensation that sits at a border, or boundary. It is not dichotomous – but rather fluid, negotiated, oscillated. For Djohari *et al.* (2018: 357) comfort is something arising from an ongoing relation with (in) the world. This is echoed by Pickerill (2015: 1065) who refers to comfort as a process, not an attribute. These world-making sensibilities have now long been attended by cultural geographers. It is precisely the politics and power which comfort makes and un-makes worlds that we humbly suggest this volume develops and contributes too.

In January 2017 Jennie Platt and her two sons decided to act on anti-homeless spikes in Manchester City Centre. They visited high-street retailer Primark to bulk buy cushions to furnish and alter these spaces. In doing so, they were attempting to make them more comfortable, enacting a quietly political act of care and compassion. It is unsurprising that attendance to comfort has thus far been largely advanced by feminist geographers interested in care, emotion, compassion and the everyday, folded into the now established debates around emotion and affect. Indeed, more recently, there has been geopolitical attention paid to the quiet, everyday, activist ways of making changes. Kye Askins (2015: 275) refers to the "*quietly political*" that is "performed through relationships that are enabled by and mutually co-productive of everyday geographies". In this light, this intervention by Jennie Platt worked to tip the everyday geographies of discomforting anti-homeless spikes back to comfort through the addition of cushions.

I am Tower of Hamlets, as I am in Tower of Hamlets, just like a lot of other people are (2011–2012) was a year-long off-site artwork by artist Amalia Pica, commissioned by Chisenhale Gallery in Bow, East London. Starting in June 2011, *I am Tower of Hamlets* involved touring a sculpture made by Pica around 52 homes and workplaces within the London Borough of Tower Hamlets. The sculpture was hand-carved by Pica out of pink granite in the form of a South American plant known for its adaptability to different environments. Any resident of Tower Hamlets could request to host the sculpture for one week. At the end of each week the current host then delivered the sculpture to the next. The sculpture's 52-week journey was bookended by two social gatherings at the gallery for the sculpture hosts.

The social encounters through this project were posited as generative, comforting and communal nature. People were brought together by the artwork through commonalities – their residence in the same Borough, their hosting of the same sculpture and, crucially, being part of the Chisenhale Gallery's art community. In this sense, the artwork did some interesting, meaningful work enhancing "affective capacit[ies]" (Thrift 2004: 128), and suggesting "an ethos…which adds to the world by framing an energetics of encounter in creative and caring ways which add to the potential for what may become" (Popke 2009: 83). However, *I am Tower of Hamlets* was a safe artwork – it involved a series of "comfortable encounters" which did not reach beyond the middle-class art community of the gallery, perpetuating problematic class-based divisions (McNally 2019). In this sense, according to some, the absence of tension and conflict – of discomfort – limits the interventional politics of the artwork (Foster 2003; Bishop 2004; McNally 2019).

Comfort then, as the illustrations above suggest, is relationally contingent to discomfort. And it is too simplistic to align the presence of comfort or discomfort with a specific politics. The chapters in this book, whilst varied in empirical and conceptual focus, each explore, investigate and interrogate the liminalities, oscillations and spatial relations between comfort and discomfort. Comfort and feeling comfortable is ordinary but at its best it is porous, generative of conviviality, generosity and empowered embodied experiences. But, if those feelings, doings, and making become "zoned" that comfort zone is dangerous. It becomes policed, bordered, a state to protect. The transformative, multiple potentials of comfort are flattened, at their worst, resulting in discomfort – of selves, of "others". This book attends to some of the "ordinary affects" emergent within this oscillation between comfort and discomfort, and opens out their "varied, surging capacities to affect and be affected that give everyday life the quality of a continual motion of relations, scenes, contingencies, and emergencies" (Stewart 2007: 1–2). Through this attention we seek to make space for the geographies of comfort to be taken seriously as a generative, transformative and evolving politics that illustrates the making and unmaking of worlds and spaces through entanglement and creativity of bodies, materials, encounters.

Bodies, identity, encounter

Since the 1990s, feminist geographers have advocated for the spatial contribution of bodies; bodies are themselves, places. Places of discourse and power relations that are mapped, embodied and resisted – where identities are performed and constructed (Rose 1993; Tolia-Kelly 2010; Bonner-Thompson and Hopkins 2017). Not only this, but bodies are spatially contingent and performative – their doings, makings and feelings are fluid and varied. Comfort is a negotiated, bodily practice. Geographers have explored how bodies negotiate feelings, atmospheres, materials and other bodies through learning and skill, for example, yoga, dance, running, painting, or knitting (Lea 2009; Price 2015; Narbed 2017; Price and Hawkins 2018). It is in these ruminations that we explore the bodily experience of comfort and how that sensation shapes being in the world. For Sara Ahmed (2007: 156),

> comfort is about an encounter between more than one body, which is the promise of a "sinking" feeling. To be comfortable is to be so at ease with one's environment that it is hard to distinguish where one's body ends and the world begins.

At the same time, comfort might also mean being aware of one's discomfort and the use of boundaries around the body to act and move through spaces (Johnson 2017: 280).

It is imperative to reaffirm that the physical, bodily sensation of comfort, of safety, in place is critically shaped by the intersections of gender, class and race. As Holliday (1999: 481) argues,

> comfort signifies the comfort one feels from the degree of fit between the outside of one's body and its inside (not blood, guts or organs, but the "imagined" or "true" self) – the way in which identity is mapped onto the body.

Such theoretical and ethnographic engagement with comfort and identity politics has been much advanced by geographers engaging with gender and sexualities. In her seminal paper on *The Comfort of Identity*, Ruth Holliday suggests

> comfort is ultimately produced in the harmony of self-explanations and self-presentations – the degree of fit between one's explanation of/for oneself and one's expression of that self – matching the inside and outside of one's body; becoming a writerly text. Where some disjuncture appears between these discourses, discomfort is produced.

Nina Held's (2015) research into Manchester's Gay Village further highlights the importance of intersections of social identities and belonging in spaces, offering a "rethinking of comfort and safety as not just feelings individuals have but as being constitutive of sexual, gender, and racial subjectivities and spaces" (2015: 32). The process of becoming comfortable in space and place is a negotiation between individuals, social groups and institutions. Perhaps even comfort "is to be feared" since it is "discomfort, displacement, disruption which moves (queer) politics (and selves) forward into a more complex and less exclusive or complacent place" (Holliday 1999: 487).

The collective understanding of what is "comfortable" can affect individual understandings of comfort, just as every individual arrives with a set of personal comfort standards that feed into the collective understanding (Cole *et al.* 2008: 333). There is a mundanity associated with comfort, it is familiar, secure, it is also to some degree the absence of change or disruption – the point that we feel comfortable with ourselves and "others". In this sense, as Yarker (2019: 535) proclaims, comfort manifests itself as a "sense of familiarity rooted in long periods of residence, safety, security and an ability to identify with those around you". Comfort encapsulates ideas such as belonging, familiarity and security. It is dangerous when people feel this is "threatened" and ascribe others as disrupting their comfort (Ahmed 2000). However, as it has been suggested, "comfort allows us to capture the affective and emotional dimension of belonging, of what belonging to a place feels like, as well as how it is practiced" (Yarker 2019: 535). The role of comfort here is key. Feelings of (dis)comfort are triggered by either perceived (or imagined) sameness/cultural familiarity or difference (Held 2015: 40). The politics of belonging, and that the process of becoming comfortable in place is a negotiation between the individual, other social groups and institution (Yarker 2019: 549). This is illustrated by Melissa Butcher and Luke Dickens in their research on young people's sense of belonging and identity through gentrification: "familiarity with places and people create a sense of comfort and security, but when disrupted this sense of belonging [is] transformed" (Butcher and Dickens 2015: 4).

In context of the global challenges of urban change, with increasingly diverse populations and mobilities, geographers have grappled with the interstices of "encounter" that comfort is to be both sought and challenged. In the first instance it is to be sought as an encounter that connects different people, in the second it is to be challenged, as the encounter must transform preconceptions (Valentine 2008; McNally 2015, 2019; Wilson 2017). The increasing "throwntogetherness" of place has meant social encounters have become distinct sites where the tension between comfort and discomfort is played out and (re)negotiated (Massey 2005). Within these encounters ideas of ethnicity, sexuality, gender, tolerance, hospitality, difference and sameness are performed, constructing and reconstructing experiences of comfort and discomfort in the process. Such encounters can be fleeting or sustained, from a distance or up close, comfortable or discomforting, but within them all lies the lived, everyday experience of living with difference. As Jackson *et al.* (2017: 2) note, "places, and indeed people, become characterised by levels of uncertainty, change, and distrust, it is crucial to think about how we negotiate our individual and collective relationships with 'the other'".

Place, environment and design

How do places *become* comforting in social and emotional terms? Comfort is a process; and the production of places that can be deemed comforting is a social, material, political and emotional process. Avril Maddrell writes that places which take on meaning in relation to the dead can "act as a catalyst, evoking grief, memories, sadness and comfort – or an unpredictable combination thereof" (2016: 170; also see Jedan *et al.* 2019). For Dewsbury and Bissell "places emerge *in habit*, through the repetition of practices and performances, itineraries and routines" (2015: 23, original emphasis). We see a relationship between this and an emergent sense of comfort in places: "Through repeated inhabitation, our sense of place can change in profound ways. As our experience tells us, the strange can become familiar; the exciting can become dull; the unseen can become perceptible" (Dewsbury and Bissell 2015: 23). Djohari *et al.* (2018: 10) discuss how comfort articulates the sense of feeling *in* a place that you cannot fully articulate why. In their research on waterscapes as places of affective geography for young people and their emotions, "comforting" spaces are continually "emerging and remade" through on-going, active interactions with(in) the world (Djohari *et al.* 2018: 4). On the contra, Kate Boyer (2012: 522) draws attention to the feeling in/out of place in the built-environment through the example of public breastfeeding and the idea that certain bodies "get in the way" (either materially, symbolically or both) disrupting the personal and public affective energies of comfort. Mothers breastfeeding often feel they must act to maintain public comfort of others by covering the bodily realities of lactation and navigating the built environment for spaces where such activity may belong (Boyer 2012: 552).

Often, comfort is used to articulate a sense of belonging in place – the scale at which people feel most "at home" (Yarker 2017: 4). The privileged and simplistic *home as haven* thesis has been much contested and decried, particularly and attentively by feminist geographers engaging with research on intimacy, violence against women and domicide (Brickell 2012a, b; Nowicki 2014; Wilkinson 2014). That said, the geographical imaginations of home conventionally demand that it is a hopeful term – a place of refuge. It is a domestic space that should be comfortable – that is, that it should be safe and secure; with a sense of familiarity afforded by a good period of residence and familiarity with those around you (Yarker 2017: 3).

The notion of home and domestic space as the archetypal *place* of comfort is powerful. It shapes research, politics, policy and social justice to contra situations where it is otherwise. For example, Darling (2011: 268) highlights ways "the mobility and insecurity associated with asylum housing, which creates discomfort and disorientation (...) acts as a means to control and constrain those awaiting decisions" (Darling 2011: 268). Indeed, these governmental processes produce "positions, techniques and places of discomfort, spaces in which feelings of belonging are undermined". Not only this, but this "politics of discomfort" acts to reassure and enable the comfort and "ease" of others who feel borders and diverse identities. In this regard then, the relationship of place with comfort is complex and nuanced; negotiations and b(ordering) of who and what it included and excluded, who belongs and who does not, and the power relations that shape such decisions – often such systems, structures and governments can be best examined on the scale of comfort – its presence, absence and the pursuit of.

Social practices are central to understanding "comfort" as a social phenomena and the relations between materials, materialities and competences of everyday life and home-making (Shove 2003; Pink 2012; Shove *et al.* 2012; Madsen 2018). Negotiating the senses and the sensory in producing an understanding of comfort is *central* to how places, dwellings and home are used, and how everyday life practiced:

> houses do not exist only as material structures but, at the same time, are homely or unhomely homes, in which the comfort of the residents might depend on physical, material, mental and social aspects connected to the idea of home.
>
> (Madsen 2018: 349)

The pervasive, historical association of home and domestic space as a site of comfort is such that the quality of "homeyness" is somewhat synonymous with comfort. As Pickerill (2015: 1065) argues, "homeyness is comforting and is created through embodied relations with materialities of the house". Negotiations of such qualities come to the fore when domestic space is imagined otherwise, and perceptions or expectations of "comfort" are challenged, for example through eco-housing. Indeed, it is possible to create comfortable spaces through alternative assemblages for example through off gridders' domestic practices that show vividly "what it means to achieve comfort differently – in variable intensities and through different technologies" (Vannini and Taggart 2014: 1078).

Enviro-comfort relations are complex and nuanced; produced and regulated by interconnected networks of global, national, local and everyday energy and energies. As Shove *et al.* (2008: 310) effuse, "comfort should be seen as permanently puzzling, an always dynamic achievement (not as something that has been definitively modelled and captured". The feeling and doing of comfort are determined through a relationship among climate, bodies and culture; people create and adapt to climate and comfort in meaningful and various ways (Fuller and Bulkeley 2013: 68; Pickerill 2015: 1064). Engaging and interrogating the intersection of historical cultural expectations of comfort with climatic conditions and everyday activities invites space to imagine how indoor and outdoor comfort might be imagined otherwise (Shove *et al.* 2008; Fuller and Bulkeley 2013: 64). Such investigation, we argue, is urgent given that in global terms, the energy and resource cost for maintaining comfort indoor and outdoor comfort has planetary consequences and implications (Chappells and Shove 2005; Shove *et al.* 2008).

More than this, historically produced western notions of comfort, for example, social and cultural expectations of the *in/out placeness* of sweat and sweating bodies are pervasive (Hitchings 2009; Waitt 2014). Thus, the design, consumption and regulation of comfort: the heating and cooling of air, water and materials is deeply entwined with global systems of spatial inequalities articulated through gender, race, class and income. As Smith and Tyszcuk (2018: 11) remind, "humanity's relationship with energy has been varied, contingent and experimental – the future of energy is ours to shape. Pressing environmental risks demand we take this task in hand". Examining our personal experiences of comfort with a commitment to understanding others allows us to understand the ordinary, everyday negotiations of comfort under climate emergency – what we accept for ourselves, and what we should *not* accept for others: migrants, refugees, displaced people whose lives are shaped by discomforting journeys in the search for social, emotional, cultural, political and environmental comfort.

Materials, materiality and makings

Ethnographic engagement with "stuff" by material geographers has long drawn attention to the emotional and affectual vibrancy of materiality, texture and things through which everyday life is transacted – from the space of the home (nik-naks, furniture, sheds, washing lines) to the space of the body – (the clothes we wear, the food we eat) (Gregson and Crewe 2003, 2011; Crang 2010; Carr and Gibson 2016; Price and Hawkins 2018). Over a decade ago, in *The Comfort of Things*, Daniel Miller (2008) illustrated the ways in which mundane, ordinary objects and materials make life meaningful through the register of comfort. These *comforting* things are collected, displayed, hoarded, disputed, discarded, repaired or displaced – and through dwelling and living with this "stuff" worlds are produced, negotiated and contested. At the same time, political theorist Jane Bennett (2010) coined the term "vital materialism" – in doing so encouraged us to engage with "the lively powers of material formations". Indeed, cultural geography has since seen a growing range of work examining the *active* role of the nonhuman in the making and unmaking of material worlds; for example geographer Divya Tolia-Kelly (2013: 154) argues for a move beyond passive "surface accounts" of matter and materiality, towards ones which understand these as "active and co-constitutive of their geographies, places, sites and space". Attending to this material *vibrancy* encourages a deeper, more critical engagement with the nonhuman things in how they actively participate in the world.

Engaging with such vibrancy then, it is true that certain materials, certain cloths, fabrics and textures are more traditionally, globally, associated with comfort. As textile historian Beverly Gordon (2011: 146) notes, "everyone, in every culture, uses and understands cloth; everyone has kinetic experience with fabric and its comforting properties". Yet seldom do textiles feature in material cultural geographic analysis (Price 2015; Stanes and Gibson 2017). Focusing on an embodied approach to materials and haptic properties of touch and feel, Stanes and Gibson (2017: 27) argue everyday geographies of clothing can be made transparent through the register of comfort/discomfort: flowy, itchy, loose, tight. At the same time, comfort is far more than the feel of a fabric, "this physical experience also encodes a sense of what seems suitable or appropriate for a particular person" (Miller and Woodward 2012: 65); "comfort is a common aspiration to a state in which people can say they feel pretty comfortable – with their clothes, social situation and themselves" (Miller and Woodward 2012: 83).

Johnson (2017: 278) argues that by focusing on the sensation of comfort we might understand how gendered, racialised, classed bodies use different clothing practices across different spaces. More than this, we might also understand how certain materials, textures and clothing become spatially located and coded as in/out of place through. Take pyjamas for example, historically conative of comfort, cosiness and conducive to sleep. Appleford (2016: 162) describes how, in the UK, being dressed in your pyjamas, in public spaces, is not deeply problematic for working-class women who are "comfortable and content with being seen in forms of private or invisible dress". Whilst for their middle-class counterparts being seen in your pyjamas is something which should be avoided, at all cost. Class distinction becomes mobilised by the sensorial, embodied comforting experience of wearing soft textured clothing and the existential *comfort* of feeling in/out of place and society.

It is not only fashion that has long been produced and adapted through *comfort* – but food too. On the corporeal experience of *eating*, Gill Valentine (1999: 334) reminds it is an enjoyable activity, a part of socialising, a hobby, or a "comforting sensation" that resonates among complex emotional feelings, including boredom and loneliness. Food that comforts, or comfort food, is produced through complex sensorial, visceral and social politics of taste, belonging and place. However, Troisi and Wright (2017: 78) suggest the popular media have oversimplified the concept of comfort food as purely unhealthy food, often consumed in moments of stress or sadness. Firstly, comfort food is an "elastic" category – it means different foods to different people; it is shaped by social, cultural, political and economic geographies. For example, in their research with migrant women in Hamilton, New Zealand, Longhurst *et al.* (2009) illustrate how taste, texture and smell help the negotiation of home. Secondly, the notion of comfort food has a complex morality – the privilege of eating purely for pleasure, placing the eater at the fore rather than considerations of sustainable food systems, supply chains or health (Jones and Long 2017). Thirdly, critical engagement with comfort food is often overlooked because of its associations with the domestic sphere. Yet, Longhurst *et al.* (2009) illustrate the visual, auditory, tactile, and skilful cooking practices of migrant women staying connected to home through familiar foods, and the importance of the kitchen as a space to produce these relations.

We reflect now, on the notion of skill and craft in the creation, maintenance and disruption of comfort. Comfort is processual; it is materially, emotionally, socially, and politically characterised by on-goingness between matter, materialities and bodies. It is fraught with tension, it is creative (Ingold 2007, 2010). This is at odds with configurations of "comfort" as conservatism, or even the space of degrowth – the comfort zone. Comfort, we argue, is a making process: "a material conversation – a physical provocation and a response, iterated over and again, working with material to understand its capacities, analyse error and make adjustments" (Carr and Gibson 2016: 303). Conceptualising comfort this way calls attention to the ways bodies, practices and spaces are enacted: the notion that comfort can be understood as an aspiration and/or achievement (Hinton 2010) has social, emotional, economic but crucially *material* capacities. When it comes to humans and our material relationships with the world and its things – ordinary comfort is imperative, it shapes decisions to stay, to move, to keep or to discard. These everyday decisions shape local, intimate human experience and response to twenty-first century global challenges such as climate emergency, migration, health and responsible consumption and production.

Investigating geographies of comfort

As Bissell (2008: 1697) argues, "comfort is a highly complex sensibility and one that requires sustained attention to the nuances therein". Geographers have begun to generate rich empirical materials on "(dis)comfort" and "(dis) comforting" experiences but, despite its colloquial prevalence as a term to understand our relationship to space and place, the disciplinary engagement with comfort has remained largely under-theorized and in need of consolidation. With this volume we have aimed to meet this need in Human Geography, providing a sustained commitment to defining, understanding and developing the "geographies of comfort". Comfort and discomfort, we argue, provide a lens through which to develop new insights on central geographical themes, including embodied relationships to environments, encounters with difference, the material textures of place, and spaces of health and wellbeing. Across the following 15 chapters, *The Geographies of Comfort* brings together recent research by scholars on these issues that endeavours to achieve three specific aims.

First is to develop the critical theorisation of comfort and its geographies. Comfort is an ambivalent and highly complex term (Bissell 2008, and this volume). To be in one's comfort zone is perceived to be conservative, and socially and culturally unadventurous (see Prezeres, this volume). At the same time the embodied, material experience of "comfort" is anticipated for satisfying experiences of everyday life. To comfort is to support and strengthen (*"confortare"*). As such, comfort is a liminal space of in-betweens and oscillations of feelings, affects and emotions (Weller 2006). Feminist geographers have argued that comfort has taken on gendered connotations, associating experiences of home, care and warmth with feminine experience and domesticity, and as such has concealed the politics, processes, making and re-making of spaces of "comfort" (Brickell 2012 a,b; see Harel, this volume). This volume works through these complex geographical issues towards a fuller understanding of the nuances, distinctions and politics of "comfort" and "discomfort".

Second is to present rich empirical research that explores the spaces, places and processes through which geographies become comforting and discomforting. Geographers and others have begun to engage with the notion of comfort in a variety of substantive contexts. This volume represents and consolidates that engagement around key cross-cutting thematics. Through this we show how comfort and discomfort are embodied geographies, implicated both in immediate environmental sensing and wider environmental cultures and politics (see Bissell et al., this volume). We investigate how comfort and discomfort modulate social encounters and constructions of difference (see Webb et al., this volume). We explore comfort and discomfort as ways to investigate the complex material textures of place (see Ward, Jayne, and Price, this volume). And we analyse the role of comfort in forging spaces of health and wellbeing (see Gorman-Murray, Martin, and Little and Morton, this volume). Throughout, the volume's contributions demonstrate how broader theoretical turns towards materialities, embodied geographies, affect and emotion are being developed through rich, innovative empirical research on geographies of comfort and discomfort.

Third, and finally, to set an agenda for future geographical engagement with comfort. This volume provides a critical engagement with the concept of "comfort" and argues for its importance in understanding our relationship to being-in-the-world. In doing so it provides a collective of chapters which each presents rich empirical data whilst developing a broader conceptual framework for geographies of comfort.

Structure of the book

With the framing of geographies of comfort in the section above we now turn to the structure of the chapters that follow. Ultimately, these chapters seek to empirically and conceptually engage with the under-addressed concept of comfort in geography. They do this, we contend, through four key themes; Bodies and Environment; Difference and Encounters; Materiality and Politics; and Health and Wellbeing, and it is these themes that we have structured the book around.

Section one: Bodies and environment

This section of the book foregrounds understanding of comfort and discomfort as embodied environmental relationships. This theme will recur throughout the volume but is explored in particular depth by the four chapters here. The section starts and ends with contributions focused on everyday embodied experience (Bissell on commuting, Adcock on going to bed / sleep). The intervening contributions from de Vet and Harel are designed to extend this focus into wider debates over our environmental relations, through attention to climate change and weather and natural disasters respectively. David Bissell carefully draws our attention to the complex, shifting and relational aspects of comfort and discomfort during commuter travel. Exploring a case study of commuting mobilities in Sydney, this chapter engages with comfort and discomfort as processual and emergent experience, and how this ebb-and-flow affects the wellbeing of urban inhabitants. Eliza De Vet addresses everyday experiences of weather in Darwin and Melbourne and how individuals navigate these in different spaces in order to seek comfort. In doing so De Vet provides a key empirical account that highlights the diversity and complexity of comfort as a mode of experience, whilst providing a critique of the use of comfort in attempts to curtail air-conditioning energy demands. Stephanie Harel's chapter focuses on the crucial role of place and home in experiences of comfort through an analysis of narratives from victims of the 2011 flood in Brisbane, Australia. From this account, Harel provides an understanding of comfort as something which "strengthens, supports and consoles in the aftermath of tragedy" (Harel, p. X, this volume). Harel does this through looking at the materiality of domestic space and the role this plays in seeking comfort in a post-disaster landscape. Finally, Jamie Adcock looks at how children's literature creates imaginary worlds which mediate between explorations of comfort and discomfort. This presents a novel geography of home, exploring how picture books both present and perform bedrooms as spaces of safety, danger, fantasy and adventure, with the aim of child development.

Section two: Difference and encounter

This section of the book foregrounds discussion of how comfort and discomfort are implicated in social encounters and negotiations of identity and difference. It starts with an analysis from Daniel Webb who draws from ethnographic research of train travel in Jerusalem to provide a critical account of the everyday geographies of the Israeli-Palestinian conflict. Deploying comfort and discomfort as a conceptual lens, Webb argues for a politics of (dis)comfort, via "indifference", which highlights distinctly geopolitical consequences. The two subsequent chapters explore the production and ambivalence of spaces of comfort for transnational populations, taking the very different cases of student mobility from Canada to the Global South (Laura Prazeres) and the Brazilian diaspora in the UK (Graça Brightwell). Prazeres' chapter addresses the crucial role of comfort in the relationship between mobility and international students' sense of home. In doing so, she describes a complex and varied account of comfort, one the one hand as something to move beyond in terms of the choice to study abroad, and the other as a feeling sought as a reaction to homesickness after moving. This chapter, then, highlights the importance of mobility regarding different registers of comfort and discomfort in people's lives. Brightwell's chapter studies the relationship between comfort, food and migration beyond the "ethnic undertones" often perpetuated in the understanding of "comfort food" (Brightwell, this collection). Drawing from primary research on diasporic Brazilian communities in London, comfort food is explored as a crucial aspect of how life in the UK is experienced and navigated. Comfort *in* food, in this regard, emerges as a confluence of material, political and emotional for migrant experiences of belonging and difference. Weiqiang Lin's final chapter in this section picks up on these themes of comfort in mobility, but pursues them in relation to the commodification of comfort for the kinetic elite of air travel, focusing in particular on the production and marketing of the Singapore Girl by Singapore Airlines. Together, the chapters consider both key conceptual terrains (such as the making of affective atmospheres) and everyday discussions about comfort and place (e.g. in relation to "comfort zones", the discomforts of homesickness, or the equations of comfort with privilege).

Section three: Materiality and texture

The materialities of comfort and discomfort are central concerns throughout the book. In this section, they are foregrounded in three chapters that focus in particular on the affordances of materials and material assemblages and their relations to the textures and politics of place. In Chapter 10, Mark Jayne develops the themes of identity and difference explored in Section 2 in an interpretation of the comfortable, "homely" dress worn in public by working class residents of Bratislava. They argue that these clothing materials are crucial to performing a material claim on urban public space and culture. Laura Price's contribution extends the textile theme, moving from consumer cultures of dress to the making cultures of knitting and the (dis)comforting qualities of wool. Finally, Miranda Ward's chapter offers a distinctly bodily insight into comfort, exploring comfort and discomfort as relational to the "correspondence" between bodies, practices and materialities in lane swimming. Drawing from rich autoethnographic research, Ward takes a close look at the *stuff* that constitutes swimming, such as tiles, chlorine and goggles, and how these converge to produce bodily experiences of watery places. Together, these chapters demonstrate how corporeal senses, emotional states and social judgments of comfort and discomfort are worked through specific material relations and affordances.

Section four: Health and wellbeing

The chapters in this section explore the emotional and psychological importance of comfort in times of death, illness and wellness. They combine attention to individual experience and agency (e.g. in Andrew Gorman Murray's exploration of the therapeutic qualities of place engagement through photography) with recognition of the institutional (in Daryl Martin's contribution focused on healthcare architecture, and in Jo Little's and Katherine Morton's chapter focused on the wellbeing industry of spas). In the form of a photo-essay, Andrew Gorman Murray used images and text to reflect on personal experiences of dis/comforting geographies at a variety of scales and spaces including the body, home, public sphere, national belonging and being-in-the-world. In doing so the chapter contributes to our understanding of the role of the visual and the *uncanny* in creating geographies of comfort. Daryl Martin's focus on healthcare architecture provides insight into the affective qualities of their buildings, and how their designs seek to enact spaces of reflection, sociability and healing. This highlights the complex emotional-affective experiences of healthcare buildings – as spaces that are familiar and unfamiliar, ordinary and extraordinary, comforting and challenging. Finally, Jo Little and Katherine Morton explore the shifting and permeable boundaries between "health" and "beauty" within contemporary constructions and performances of femininity. Exploring the spa as a space of bodily repair, maintenance and comfort, it suggests how routine practices of relaxation and pampering engaged in by spa clients incorporate strong expectations of bodily management and discipline. Together, these contributions reflect on the ambivalent association of comfort with technologies of self-management and on the potential personal value of engaging directly with spaces of comfort and discomfort.

Towards geographies of comfort

In Lauren Berlant's (2011) *Cruel Optimism* she powerfully enunciates the harmful side of comfort, where the ordinariness of crisis in contemporary life becomes both entrenched and tempered by fantasies of achieving a comfortable "good life". Comfort, in this form, is "the attrition of a fantasy" (Berlant 2011: 11). Why then, should we engage with comfort? We worried that a focus on "comfort" would belie the complex, urgent, political geographies of (dis)comfort – that our engagement with comfort would privilege voices, geographies and uphold narratives of those for whom life is indeed, *comfortable*. Indeed, as Holliday (1999: 489) reminds us, comfort is "an easy, unthinking state (…) being comfortable – as in comfortably off – implies a lack of necessity to worry about the world or one's position in it" (Holliday 1999: 489). Our agenda, rather, is to illuminate the importance of engaging with comfort as a world-making activity and to encourage geographical engagement with this as such, not only to add to the long list of emotions for which geographers interrogate but as a sensibility that is so *unthinking* and thus requires deep engagement with in order to investigate the familiar and imagine worlds otherwise.

As Skeggs *et al.* put it, "comfort is defined against a wider experience of danger and insecurity in regard to physical violence, in contrast to a more diffuse form of threat or a wider spectrum of insecurity, danger, and loss of safety" (2004: 84). For decades, geographers have been critically investigating precarity and in/security at a structural and systemic level – informal labour, housings, and educational and environmental relations (Philo 2012). At any one time the line between security and insecurity – what is secure and what is not – looks different for differently located people (Noxolo 2017: 6). Geographers should pay careful, critical attention to the ways in which boundaries are drawn when people are trying to make themselves feel secure – and how this is negotiated in everyday lives (Noxolo 2017: 6). By exploring the geographies of comfort might we connect the impact of the political, and security of bodies, identity and place with the politics and security of ordinary, everyday lives?

Through its association with domestic space and home – comfort, we argue, has been overlooked as unimportant, trivial and mundane. As Susan Fraiman (2017) argues, "domesticity gets a bad rap. We associate it with stasis, bourgeois accumulation, banality, and conservative family values" – yet it is just as likely to involve dislocation, economic insecurity, creative improvisation and queered notions of family. Such bodies, spaces and practices are precisely the location of intimate critical feminist cultural geographic thought and research (Brickell 2012a, 2012b; Pain 2015; Noxolo 2017). As Fraiman (2017: 166) puts it, locating discomfort and domestic struggle reaffirms respect for "domesticity as security, the importance of warm, tactile connections, the sense of belonging and obligation". This collection is indebted to the work of feminist geographers, and it bears repeating that women, globally, continue to inordinately *feel* the effects of precarity and in/security in everyday life – and yet at the same time labours of care, and doing, or un-doing comfort, more often than not, sit with women. For example, Sara Ahmed's (2010) "feminist killjoy" who disrupts everyday life and public comfort through the calling out racism or sexism. Examining comfort and comfort zones is by definition an uncomfortable process – but it could not be more important in the quest to make public space familiar, welcoming, transgressive and shared (Zebracki 2016).

This collection seeks to illustrate the ways that humans seek comfort and their geographies. How worlds are actively made and unmade processually in pursuit of comfort. In doing so, it highlights the shared importance of bodies, materials and practices in producing space, place and institutions. Though comfort, and the comfort zone especially, has historically implied lack of mobility we argue that comfort is rather a liminal space: an oscillation. The process to produce, maintain or describe comfort is one of creativity, skill and tension – it is *world-making*. We should engage critically with who does this work, where it takes place, whose experiences are privileged, how comfort is materialised, actioned and verbalised, and the times, spaces and place that discomfort becomes comfort. Human experience at its best should be described as *comfortable* and by paying attention to the intimate, personal, ordinary worlds through which this is achieved, negated, or made precarious and temporary is to understand the complex worlds of humans, homes and the stuff that brings them to life.

References

Adey, P., Bissell, D., McCormack, D.P. and Merriman, P. (2012) 'Profiling the Passenger: Mobilities, Identities, Embodiments.' *Cultural Geographies*, 19(2), 169–193.

Ahmed, S. (2000) *Strange Encounters: Embodied Others in Post-coloniality*. Abingdon: Routledge.

Ahmed, S. (2007) 'A Phenomenology of Whiteness.' *Feminist Theory*, 8(2), 149–168.

Appleford, K. (2016) 'Being Seen in Your Pyjamas: The Relationship between Fashion, Class, Gender and Space.' *Gender, Place & Culture*, 23(2), 162–180.

Askins, K. (2015) 'Being Together: Everyday Geographies and the Quiet Politics of Belonging'. *ACME: An International Journal for Critical Geographies*, 14(2), 470–478.
Bennett, J. (2010) *Vibrant Matter: A Political Ecology of Things*. Durham: Duke University Press.
Berlant, L. (2011) *Cruel Optimism*. Durham: Duke University Press.
Bishop, C. (2004) 'Antagonism and Relational Aesthetics.' *October*, 110, 51–79.
Bissell, D. (2008) 'Comfortable Bodies: Sedentary Affects.' *Environment and Planning A: Economy and Space*, 40(7), 1697–1712.
Bonner-Thompson, C. and Hopkins, P. (2017) *Geographies of the Body*. New York: Oxford University Press.
Boyer, K. (2012) 'Affect, Corporeality and the Limits of Belonging: Breastfeeding in Public in the Contemporary UK.' *Health and Place*, 18(3), 552–560.
Brickell, K. (2012a) 'Geopolitics of Home.' *Geography Compass*, 6(1), 575–588.
Brickell, K. (2012b) '"Mapping" and "Doing" Critical Geographies of Home.' *Progress in Human Geography*, 36(2), 225–244.
Brighenti, A. M. and Pavoni, A. (2019) 'City of Unpleasant Feelings. Stress, Comfort and Animosity in Urban Life.' *Social & Cultural Geography*, 20(2), 137–156.
Butcher, M. and Dickens, L. (2015) *Creating Hackney as Home*. Available at http://www.hackneyashome.co.uk/sites/default/files/documents/Creating_Hackney_as_Home_Final_Report.pdf. Accessed 31 October 2019.
Carr, C. and Gibson, C. (2016) 'Geographies of Making: Rethinking Materials and Skills for Volatile Futures.' *Progress in Human Geography*, 40(3), 297–315.
Cole, R. J., Robinson, J., Brown, Z. and O'shea, M. (2008) 'Re-contextualizing the Notion of Comfort.' *Building Research & Information*, 36(4), 323–336.
Colls, R. (2012) 'Feminism, Bodily Difference and Non-representational Geographies.' *Transactions of the Institute of British Geographers*, 37(3), 430–445.
Crang, P. (2010) 'Cultural Geography: After a Fashion.' *Cultural Geographies*, 17(2), 191–201.
Crewe, L. (2011) 'Life Itemised: Lists, Loss, Unexpected Significance, and the Enduring Geographies of Discard.' *Environment and Planning D*, 29(1), 27–46.
Crouch, D. (1989) 'The Allotment, Landscape and Locality: Ways of Seeing Landscape and Culture.' *Area*, 21(3), 261–267.
Crowley, J. E. (1999) 'The Sensibility of Comfort.' *The American Historical Review*, 104(3), 749–782.
Darling, J. (2011) 'Domopolitics, Governmentality and the Regulation of Asylum Accommodation.' *Political Geography*, 30(5), 263–271.
Dewsbury, J. D., and Bissell, D. (2015) 'Habit Geographies: The Perilous Zones in the Life of the Individual.' *Cultural Geographies*, 22(1), 21–28.
Djohari, N., Brown, A., and Stolk, P. (2018) 'The Comfort of the River: Understanding the Affective Geographies of Angling Waterscapes in Young People's Coping Practices.' *Children's Geographies*, 16(4), 356–367.
Eden, S. (2014) 'Out of the Comfort Zone: Enhancing Work-based Learning about Employability through Student Reflection on Work Placements.' *Journal of Geography in Higher Education*, 38(2), 266–276.
Eldridge, A. and Roberts, M. (2008) 'A Comfortable Night Out? Alcohol, Drunkenness and Inclusive Town Centres.' *Area*, 40(3), 365–374.
Foster, H. (2003) 'Arty Party.' *London Review of Books*, 25(23), 21–22.

Fraiman, S. (2017) *Extreme Domesticity: A View from the Margins.* New York: Columbia University Press.
Fuller, S. and Bulkeley, H. (2013) 'Changing Countries, Changing Climates: Achieving Thermal Comfort through Adaptation in Everyday Activities.' *Area*, 45, 63–69.
Gordon, B. (2011) *Textiles: The Whole Story.* London: Thames and Hudson.
Gorman-Murray, A. (2012) 'Queer Politics at Home: Gay Men's Management of the Public/Private Boundary.' *New Zealand Geographer*, 28(2), 111–120.
Gregson, N. and Crewe, L. (2003) *Second-hand Cultures.* London: Berg.
Held, N. (2015) 'Comfortable and Safe Spaces? Gender, Sexuality and "Race" in Night-time Leisure Spaces.' *Emotion, Space and Society*, 14, 33–42.
HESA (2020) *What Are Their Employment Conditions?* Higher Education Statistics Agency. Available at: https://www.hesa.ac.uk/data-and-analysis/staff/employment-conditions. Accessed 15/05/2020.
Hinton, E. (2010) *Review of the Literature Relating to Comfort Practices and Socio-technical Systems.* Environment, Politics and Development Working Paper Series, 35.
Hitchings, R. (2009) 'Studying Thermal Comfort in Context.' *Building Research & Information*, 37(1), 89–94.
Hitchings, R., Surrage, R. and Atkinson, W. (2014) 'Human Activities at the Frontiers of Climate Control: Thermal Comfort and How Shoppers and Sport Spectators in the UK Currently Talk about Air-conditioning.' *Geoforum*, 54, 103–110.
Holliday, R. (1999) 'The Comfort of Identity.' *Sexualities*, 2(4), 475–491.
Holmes, H. and Hall, S. M. (eds.) (2020) *Mundane Methods: Innovative Ways to Research the Everyday.* Manchester: Manchester University Press.
Ingold, T. (2007) *Lines: A Brief History.* London: Routledge.
Ingold, T. (2010) 'Footprints through the Weather-world: Walking, Breathing, Knowing.' *Journal of the Royal Anthropological Institute*, 16, 121–139.
Jackson, L., Harris, C. and Valentine, G. (2017) 'Rethinking Concepts of the Strange and the Stranger.' *Social & Cultural Geography*, 18(1), 1–15.
Jedan, C., Maddrell, A. and Venbrux, E. (2019) *Consolationscapes in the Face of Loss: Grief and Consolation in Space and Time.* London: Routledge.
Johnson, A. (2017) 'Getting Comfortable to Fat Home: Clothing Practices of Black Muslim Women in Britain.' *Gender, Place & Culture*, 24(2), 274–287.
Jones, M. O. and Long, L. M (eds.) (2017) *Comfort Food: Meaning and Memories.* Jackson: University of Mississippi Press.
Laurier, E. and Philo, C. (2006) 'Possible Geographies: A Passing Encounter in a Café.' *Area*, 38, 353–363.
Lea, J. (2009) 'Liberation or Limitation? Understanding Iyengar Yoga as a Practice of the Self.' *Body & Society*, 15(3), 71–92.
Longhurst, R., Johnston, L. and Ho, E. (2009) 'A Visceral Approach: Cooking "At Home" with Migrant Women in Hamilton, New Zealand.' *Transactions of the Institute of British Geographers*, 34(3), 333–345.
Lorimer, H. (2005) 'Cultural Geography: The Busyness of Being "More-than-Representational".' *Progress in Human Geography*, 29(1), 83–94.
Maddrell, A. (2016) 'Mapping Grief. A Conceptual Framework for Understanding the Spatial Dimensions of Bereavement, Mourning and Remembrance.' *Social & Cultural Geography*, 17(2), 166–188.
Madsen, L V. (2018) 'Materialities Shape Practices and Notions of Comfort in Everyday Life.' *Building Research & Information*, 46(1), 71–82.
Massey, D. (2005) *For Space.* London: Sage.

McNally, D. (2015) 'Comforting Others: Sociality and the Ethical Aesthetics of Being-together.' In Hawkins, H. and Straughan, E. (eds.), *Geographical Aesthetics: Imagining Space, Staging Encounters*. London: Routledge, pp. 121–136.
McNally, D. (2019) '"I am Tower of Hamlets": Enchanted Encounters and the Limit to Art's Connectivity.' *Social & Cultural Geography*, 20(2), 198–221.
Miller, D. (2008) *The Comfort of Things*. Cambridge: Polity.
Miller, D. (2010) *Stuff*. Cambridge: Polity.
Miller, D. and Woodward, S. (2012) *Blue Jeans: The Art of the Ordinary*. London: University of California Press.
Mountz, A., Bonds, A., Mansfield, B., Loyd, J., Hyndman, J., Walton-Roberts, M., Basu, R., Whitson, R., Hawkins, R., Hamilton, T. and Curran, W. (2015) 'For Slow Scholarship: A Feminist Politics of Resistance through Collective Action in the Neoliberal University.' *ACME: An International Journal for Critical Geographies*, 14(4), 1235–1259.
Narbed, S. (2017) *Cultural Geographies of Contemporary Dance in Quito, Ecuador*, Ph.D., Royal Holloway, University of London.
Noxolo, P. (2017) 'In/security: Global Geographies of a Troubled Everyday.' *Geography*, 102, 5–9.
Nowicki, M. (2014) 'Rethinking Domicide: Towards an Expanded Critical Geography of Home.' *Geography Compass*, 8, 785–795.
Philo, C. (2012) 'Security of Geography/Geography of Security.' *Transactions of the Institute of British Geographers*, 37(1), 1–7.
Pain, R. (2015) 'Intimate War.' *Political Geography*, 44, 64–73.
Pickerill, J. (2015) 'Cold Comfort? Reconceiving the Practices of Bathing in British Self-build Eco-homes.' *Annals of the Association of American Geographers*, 105(5), 1061–1077.
Pink, S. (2012) *Situating Everyday Life: Practices and Places*. London: Sage Publications.
Popke, J. (2009) 'Geography and Ethics: Non-representational Encounters, Collective Responsibility and Economic Difference.' *Progress in Human Geography*, 33(1), 81–90.
Price, L. (2015) 'Knitting and the City.' *Geography Compass*, 9, 81–95.
Price, L. and Hawkins, H. (eds.) (2018) *Geographies of Making*. London: Routledge.
Racz, I. (2015) *Art and the Home: Comfort, Alienation and the Everyday*. London: IB Tauris.
Rishbeth, C. and Rogaly, B. (2018) 'Sitting Outside: Conviviality, Self-care and the Design of Benches in Urban Public Space.' *Transactions of the Institute of British Geographers*, 43, 284–298.
Rose, G. (1993) *Feminism and Geography: The Limits of Geographical Knowledge*. London: Wiley.
Shove, E. (2003) *Comfort, Cleanliness and Convenience: The Social Organization of Normality*. London: Berg.
Chappells, H. and Shove, E. (2005) 'Debating the Future of Comfort: Environmental Sustainability, Energy Consumption and the Indoor Environment.' *Building Research & Information*, 33(1), 32–40.
Shove, E., Pantzar, M. and Watson, M. (2012) *The Dynamics of Social Practice: Everyday Life and How It Changes*. London: Sage.
Shove, E., Chappells, H., Lutzenhiser, L. and Hackett, B. (2008) 'Comfort in a Lower Carbon Society.' *Building Research and Information*, 36(4), 307–311.

Skeggs, B., Moran, L., Tyrer, P. and Corteen, K. (2004) *Sexuality and the Politics of Violence and Safety*. London and New York: Routledge.

Smith, J. and Tyszczuk, R. (eds.) (2018) *Energetic: Exploring the Past, Present and Future of Energy*. Cambridge: Shed.

Spinney, J., Green, N., Burningham, K., Cooper, G. and Uzzell, D. (2012) 'Are We Sitting Comfortably? Domestic Imaginaries, Laptop Practices, and Energy Use.' *Environment and Planning A*, 44(11), 2629–2645.

Stanes, E. and Gibson, C. (2017) 'Materials That Linger: An Embodied Geography of Polyester Clothes.' *Geoforum*, 85, 27–36.

Stewart, K. (2007) *Ordinary Affects*. Durham and London: Duke University Press.

Thrift, N. (2004) 'Performance and Performativity: A Geography of Unknown Lands.' In Duncan, J., Johnson, N. and Schein, R. (eds.), *A Companion to Cultural Geography*. Malden: Blackwell, pp. 121–136.

Tolia-Kelly, D. P. (2010) 'The Geographies of Cultural Geography I: Identities, Bodies and Race.' *Progress in Human Geography*, 34(3), 358–367.

Tolia-Kelly, D. P. (2013) 'The Geographies of Cultural Geography III: Material Geographies, Vibrant Matter and Risking Surface Geographies.' *Cultural Geographies*, 37(1), 153–160.

Troisi, J. D. and Wright, J. W. C. (2017) 'Comfort Food: Nourishing Our Collective Stomachs and Our Collective Minds.' *Teaching of Psychology*, 44(1), 78–84.

Valentine, G. (1999) A Corporeal Geography of Consumption. *Environment and Planning D: Society and Space*, 17(3), 329–351. https://doi.org/10.1068/d170329

Valentine, G. (2008) 'Living with Difference: Reflections on Geographies of Encounter.' *Progress in Human Geography*, 32(3), 323–337.

Vannini, P. and Taggart. J. (2013) 'Domestic Lighting and the Off-grid Quest for Visual Comfort.' *Environment and Planning D: Society and Space*, 31(6), 1076–1090.

Vannini, P. and Taggart, J. (2014) 'Making Sense of Domestic Warmth: Affect, Involvement, and Thermoception in O-grid Homes." *Body and Society*, 20(1), 61–84.

Waitt, G. (2014) 'Bodies That Sweat: The Affective Responses of Young Women in Wollongong, New South Wales, Australia.' *Gender, Place & Culture*, 21(6), 666–682.

Weller, S. (2006) 'Situating (Young) Teenagers in Geographies of Children and Youth.' *Children's Geographies*, 4(1), 97–108.

Wilkinson, E. (2014) 'Single People's Geographies of Home: Intimacy and Friendship beyond 'the Family'. *Environment and Planning A*, 46(10), 2452–2468.

Wilson, H. F. (2017) 'On Geography and Encounter: Bodies, Borders, and Difference.' *Progress in Human Geography*, 41(4), 451–471.

Yarker, S. (2019) 'Reconceptualising Comfort as Part of Local Belonging: The Use of Confidence, Commitment and Irony.' *Social & Cultural Geography*, 20(4), 534–550.

Zebracki, M. (2016) 'The Search for Publics: Challenging Comfort Zones in the Co-Creation of Public Art.' *Cultural Geographies*, 23(4), 739–744.

Section one

Bodies and environments

2 Transitioning comforts

Bodily evaluations of urban mobilities

David Bissell

In 2011, the Auditor-General of the state of New South Wales in Australia made a statement about the experience of commuting in Sydney that seemed to capture the collective sentiment of many Sydneysiders. In an interview based on the release of new figures showing that morning peak hour traffic had become worse on six out of the seven major roads in Sydney, Peter Achterstraat said that it is vital that we

> help people get to work on time and with less stress. If you've spent the morning fighting in the traffic and you feel like you've had 15 rounds with Muhammad Ali before you get to work, you're not going to be as productive as someone who is fresh.
>
> (Saulwick 2011)

Such a charged remark echoed a growing discourse about the discomforts of commuting in Sydney testified through regular column inches in the local media and high-profile reports that have sought to measure and "expose" the city's high level of commuter pain (IBM 2011). Whilst Achterstraat's statement links commuter discomfort to the risk that it poses to the economic productivity of the city, this issue clearly has much wider implications for evaluating how the experience of commuting is affecting the wellbeing of urban inhabitants.

Many surveys about commuting claim to have pinpointed what causes discomfort (for instance IBM 2011). As such, the value of posing more conceptual questions about the constitution of discomfort might not seem immediately obvious. If it is assumed that the experience of congestion and stop-start traffic for car drivers are some of the main causes of commuter discomfort, then it would seem plausible that improving infrastructure by decongesting roads and making transport networks more efficient would alleviate these problems, with little need for further questioning. Yet beyond these familiar sequences of practical problems and respective infrastructure solutions, we still know very little about what exactly commuter discomfort itself is and the effect that it is having on people who commute in Sydney. Put simply commuter discomfort is usually understood rather narrowly as a *residual effect* of poor infrastructure, rather than an ongoing experience of a wider series of relations *that generates its own effects*. Moving beyond discourses that generalise the commuting experience, this chapter seeks to remedy some of these shortcomings by exploring how commuting discomfort can materialise and manifest in different ways. It shows how comfort and discomfort are bodily evaluations of experience that change over time, often in subtle ways, in relation to changing circumstances. It demonstrates how these bodily evaluations give rise to different tactics that aim to maximise people's powers of acting.

The project on which this chapter is based is examining how practices of commuting in Sydney are actively involved in transforming the very nature of urban life. To understand how practices of commuting are changing home lives, work lives and family lives, the project is exploring some of the new experiences, socialities and ways of living that commuting gives rise to. Conceptually, to do this, the project aims to provide an account of commuting in Sydney that is attuned to the specificity and singularity of the scene that it surveys. It takes its theoretical cue from poststructuralist theories in cultural geography and mobilities research that are concerned with questions regarding processes and relations, rather than questions of identities and essences (Anderson and Harrison 2010). These conceptual perspectives enable a shift from a preoccupation with similarities and correlations to a preoccupation with how difference actually occurs. In this regard, the purpose of examining commuting mobilities in Sydney is not to construct a top-down model that can then be applied to understand commuting problems in other cities, but rather more modestly it is to draw out the distinctive relations that make up experiences of commuting in Sydney. Poststructuralist theories provide precisely this lure for enabling the singular inchoate, vague and therefore often overlooked events, encounters and moments to be drawn out and considered on their own terms, rather than falling to the "bad politics of strong theory's tendency to beat its objects into submission to its dreamy arguments", as Stewart (2013: 284) puts it.

Practically this project has involved undertaking fieldwork with commuters in Sydney. In this chapter I draw on one part of this fieldwork which is based on 53 interviews that I conducted with commuters in 2013. These commuters responded to an advertisement in the *Sydney Morning Herald* and *MX*, a free commuter paper, which called for participants whose lives were significantly impacted on by the practice of commuting. Within qualitative methods debates, interviews are often understood in a representational sense, as offering a window onto past events. However, poststructuralist theories have taught us to be wary of such claims because they overlook the way that interview encounters are themselves involved in the ongoing constitution of events. Respecting this, the interviews conducted for this project did not set out to gain a view onto an external reality. Instead, as sites of heightened receptivity to the stressful and uncomfortable aspects of commuting, interviews were conducted to explore how these discomforts and stresses were narrated and reflected upon by participants. As such, they provided me with an opportunity to draw out some of the wider webs of relations that the commute opens out onto.

This chapter is split into three parts. In the first part, I argue that within research on mobilities, comfort is often treated as a predictable effect of different technologies of transit. The problem with this is that comfort is treated in a determinist way that pays little attention to the vicissitudes of experience and the wider webs of relations that commuting is a part of. Using Sara Ahmed's writings as a point of orientation, I move forward by showing how comfort can be understood in a much more differentiated way. Pushing this focus on differentiation, in the second part, I describe two interview encounters. The aim here is to draw out the processual aspects of how comforts and discomforts were talked about and reflected on during the two interviews in distinct and contrasting ways. Reflecting on these reflections, part three highlights some practical implications for our understanding of comfort and discomfort that come out of these two interviews. Engaging with Deleuze's writings on Spinoza's ethics, the discussion moves forward to consider how evaluations of comfort and discomfort might be understood from the perspective of active and passive affections. This conceptual lens understands comfort in a much more transitional way because it is attentive to how our bodily powers of acting change through different circumstances.

Knowing comfort: from aggregates to differentiation

What counts for a comfortable or uncomfortable commute is often assumed to be already known. For instance, standing on an overcrowded train for a long time; sitting next to someone on a bus with loud headphones; waiting in a traffic jam; and getting soaked by passing cars splashing through puddles are all scenarios that immediately evoke a sense of discomfort. Such universalised assumptions about what makes for a comfortable commute are at play in the design of commuting spaces. From waiting rooms to train interiors, expert design knowledges have been used to produce segregated and sorted mobilities where different capacities to pay translate into different degrees of comfort. Research that has looked at the material design of technologies of transit illustrates this well. Jain's (2011) ethnographic research on a premium scheduled coach service between Oxford and London and my own research (Bissell 2008) on the material differences between first class and standard class train travel both show how elements of the travelling environment are made more comfortable to justify the higher fares for more elite travelling environments. Research on air travel similarly illustrates this tendency perhaps most strikingly where subtle and multiple gradations of eliteness translate into different degrees of comfort (Lin, this collection).

Reflecting this understanding, Dant (2014: 372) suggests that comfort might even be understood as a form of mobility capital. This argument is particularly convincing in relation to Richardson and Jensen's (2008) discussion of how the premium Skytrain in Bangkok gives people who can afford to use the service a comfortable and speedy ride above the more uncomfortable, slower and cheaper modes of transit below. This anaesthetisation of discomfort for those who can afford to pay parallels Graham's (2002) discussion of the Paris-Roissy airport shuttle train where travellers are shielded from the potentially less-comfortable aspects of the places that the line traverses. However, the problem with this relationship where the degree of comfort corresponds to the eliteness of the transit environment is that it sets up a probabilistic notion of comfort that overlooks the complexity of experience. In short it assumes that people will experience these transit environments in the same way, rather than appreciating how the sheer diversity of people's previous experiences will prime expectations and responses in different ways.

Sara Ahmed's writings on comfort provide us with an alternative to these universalised understandings, seeking instead to show how comfort is differentiated. Rather than speculating on how different degrees of comfort might be engineered into spaces by experts, Ahmed argues that comfort is a much more relational experience. From the point of view of sexuality, she examines how some people come to feel more comfortable at the expense of others. As such, she is interested in understanding how comfort is intimately connected to conditions of normativity, where "normativity is comfortable for those who can inhabit it" (2004: 147). She argues that heteronormativity functions "as a form of public comfort by allowing bodies to extend into spaces that have already taken their shape" (2004: 148). Comfort here is about a seamless inhabitation of the world, where one is "so at ease with one's environment that it is hard to distinguish where one body ends and the world begins" (2004: 148). For Ahmed, the power of normativity concerns how some bodies register in the urban landscape whilst others do not. As she stresses "the surfaces of the social as well as bodily space 'record' the repetition of acts, and the passing by of some bodies and not others" (2004: 148).

Ahmed is interested in comfort because of its relationship to ease. She says that comfort "suggests wellbeing and satisfaction, but it also suggests an ease and an easiness" (2004: 147). Because heteronormativity produces a comfort zone where some can inhabit space with greater ease than others, this creates inequalities that are registered through feelings of discomfort. The political consequence of this is that the comforts of heteronormativity are necessarily contingent on the discomfort of others who do not conform to such ideals. From this perspective, discomfort involves a "failure to fit" (2012: 41). As such, rather than an engineered aspect of designed environments, Ahmed's writings highlight how comfort and discomfort are much more socially-produced experiences. In particular, her work shows how the comforts and discomforts of fitting in or not fitting in can be evaluated according to specific axes of identity. Here it is specific bodies in terms of their gestures or appearance that are the cause of discomfort, or are made to feel the discomforts of not fitting in.

Yet Ahmed's focus on normativity as the basis for evaluating comfort raises some important questions for the present investigation into commuting. First, rather than just focusing on bodily encounters with other human bodies, how might the concept of comfort be broadened to account for material encounters with other dimensions of travelling environments? Second, and relatedly, rather than focusing on identity as the basis for differentiation, how might comfort come to be differentiated in a more immanent way, shaped and evaluated through the myriad ongoing encounters that bodies have? Third, how might an understanding of comfort that is differentiated through experience help us to appreciate how commuting practices come to be evaluated by bodies in much more fluid ways? Fourth, what are the practical and conceptual implications for a more transitional and less universal understanding of comfort? To respond to these provocations, in the next part of this chapter I describe two interview encounters with commuters in Sydney.

Comfort transitions

Helen's interview

I am in a windowless strip-lit room sitting on a soft low chair. Facing me, sat on an identical low chair is a woman whose situation I've got to know a bit about by email. We've been trying to tee up a good time for a meeting for about a month now that fits around her annual leave. This week Helen[1] has just returned to work after being away on holiday for her "rejuvenation" period, as she puts it. I'm looking forward to talking to her because a few weeks ago in her second email to me she was very keen on sharing her account:

> I am more than happy to, and would very much like to be, involved in your study as I think I would make an interesting subject as travelling in Sydney takes a massive amount from me each day—I have a 20% permanent impairment and also, have been diagnosed with [nerve disorder]. Having said this, I am still managing at this point in time, to come work 5 days however it's at a massive cost to me socially and mentally as I am exhausted from travelling each day.

I know from her initial email responding to the project's advert in the *Sydney Morning Herald* that she travels to work in the city's CBD by bus from a suburb in the north-west of the city. She tells me that it takes about an hour to travel each way.

To start the conversation, I ask her about today's commute. She says that there are four parts to her commute "you know; that's a lot". "It is", I agree. "Yeah; and—and that really physically kills me. And that's why I'm saying, in the morning, I'm just struggling, if I don't have myself regimented."

I ask her how long she has been doing this commute for. She tells me that she started about five years ago and at that time took the train, driving to her nearest station then catching the express service to Central. But then a few years into that commute, she tells me that her health had deteriorated. "I was extremely worn down—I thought, 'I can't cope with driving and with the train; there must be another alternative.'" I was intrigued at how her changing health conditions had prompted her to think about other travel options. She tells me how she was looking for a route that was easier on her body:

> I knew about the buses on the M2 directly, but I didn't know about the buses up the top on [name] Road. That's when I decided I was going to catch the bus because it's a far lot easier, because I've only got to drive up to the top of the street and the bus is there. I don't have to deal with getting down [name] Road, across to [place], parking my car. It's—it's a much shorter, concise, tighter journey for me, in terms of physical movement.

I respond by asking, "you made that decision based on cutting out some of the sort of excess movements?" She agrees but emphasises that it is the aspect of slowing that is important:

> Oh, definitely; most definitely... and that's—what I was trying to slow down was my physical movement, because that really does take a toll on me—not so much in the morning, because I'm still half asleep, but in the afternoons, when I'm really tired and fatigued, you know.

I'm interested to hear more about the journey itself. She tells me how her physical pain changes over time but that it "is good at the moment". She talks about permanent nerve damage that she has sustained in an accident and how this means that she has developed a sensitivity to which parts of her body can withstand pressure, particularly in terms of sleeping. I ask her whether this sensitivity has implications for how she travels. To my surprise, instead of focusing on the muscular aspects of sitting, she shifts to telling me about her heightened sensitivity to smells.

> What drives me 100 per cent now to where I sit on the bus—and you're going to laugh when I say this—is men's cologne or highly scented perfumes. Part of my [nerve condition]—and I'll be honest with you: you've got something on that's aggravating me—is heightened senses, which is why I have the glasses, because I can't have the glare and the bright lights; the smells—certain smells or certain fragrances or certain oils and the hearing: I can't—I can't—that's something that's significantly changed in my life; as someone that used to have constant noise around them, I now just live for peace and quiet.

I suddenly feel awful that my liberally-applied aftershave is having such an effect here right now. She continues by drawing my attention to how the process of boarding the bus involves becoming sensitised to where these potential discomforts might be located:

> So, as soon as I get into a bus, my nose is up because, if there's something really strong on the bus, I can't sit near it; I've got to get away from it as much as possible because it can prompt two—two things: one, vomiting; or a migraine. So it's more that than—than my physical wellbeing at that point in time.

I hope that the air conditioning in this room is going to avert these here, as she tells me, "there's been a couple of mornings, when the buses have smelt so bad with fragrances, I've got off the bus because I just know I'm going to be as sick as a dog."

Helen continues by talking about her posture on the bus. In particular, she flags how her choice of seat is partly informed by a sensitivity to preserving the capacity of her right hand and arm. She describes this in terms of her capacity to respond to sudden movements:

> I don't like to sit in the buses where there's nothing in front of me because, you know, you get those crazy drivers that do jolt you. I—I don't want to be moved forward so that I've got to put an arm out, because I—I—I won't go there.

This is clearly a tactic that threads through her work and home life too:

> And I also won't stand on a bus—I don't care how packed the—or how late I am for work—because I can't hold with my right arm. I—because I'm right-handed, I won't do anything with my right arm that jeopardises my shoulder, to the basic things of I don't lift anything probably more than a ream of paper at work. I—I won't even lift a two-litre bottle of milk at home; it goes into a jug and I use that—because I have to protect my right arm because it's my work. And I've had times where I've been in so much pain I haven't been able to work, and I don't want to do that—touch wood—again. I haven't—my arm's been okay, but I think that's because I'm quite anal about what I do and don't do with it.

She then clarifies that never putting her arm up to hold anything is "a protection mechanism to—to make sure that, you know, (a) I'm not setting myself up for an injury in some way or (b) I'm not aggravating myself". Later on in the interview, when we are talking about reading on the bus, Helen says that training herself to not put her arms up to preserve her hands for work, leaves her feeling vulnerable because her reflex actions have changed:

> As long as the driver doesn't drive like a lunatic, I'll be all right; I won't go falling down the front of the bus'—because that's my biggest fear, because my reflexes will be that I won't put out my arms and I'll land on my face, so—and that's what's happened when I've hurt myself before.

Helen's reflections on how she adapts her journey to and from work based on an attunement to her changing bodily conditions suggests that a comfortable commute is not something that is fixed. A little later during the interview, for instance, she explicitly describes this process when she says "you just learn to adapt to—to certain things." On returning to work after her accident, this adaptation involved travelling to work outside peak commuting hours, arriving at work at ten o'clock to make the journey easier, and catching taxis between the station and work because she says "I didn't want to get on buses". Later on, it involved using a walking stick for both physical assistance and visibility. And now she gets a new bus that she's discovered to minimise physical movement. Interestingly, however, she contrasts these changes with a different propensity for disliking change now. Quite vehemently, she says:

> I don't like change. Having said that, I love change—I used to love change; I used to thrive on change at work and outside of work. I was never home before my accident; I was always out and about, very, very social. Since my accident, I'm very withdrawn.

For Helen, then, it seems that there has been a long period of gradually becoming more comfortable with a changing set of bodily capacities, where she has had to learn new ways of inhabiting commuting environments. Comfort here is achieved through knowledge of her capacities, and knowledge of what causes discomfort. Indeed towards the end of the interview she relates to the concept of comfort more directly:

> When I'm very comfortable with my environment—and that comes back to the bus, doesn't it: like, knowing where I can sit; knowing my limitations when I'm moving around so that I don't bump into things or I don't bump into people. And that's—that's purely been since the accident and having a disability, you know. Whereas most people are rushing to get somewhere and sit down, no, I've got to really take my time.

Kate's interview

It is lunch time in the CBD of Sydney one warm, sunny Thursday afternoon in late February 2013. I am standing in the lobby of a large financial corporation finished with dark grey high-gloss floor tiles and muted wood panelling. As I wait for Kate, my interviewee, to meet me my pre-interview nerves are heightened by the sheer volume of smartly dressed people being released from the line of lifts. Right on time, a young woman's eyes meet mine; she smiles and walks over to me. We shake hands, exchange pleasantries after which she guides me to one of the lifts. We travel to one of the higher floors of the building, walk along a warmly lit corridor with frosted glass offices on either side to a large meeting room. She invites me to take a seat at the large table and as I sit, I quickly snatch a glance out of the large window. It is an incredible sight across to the south-west, the elevation giving a real sense of the sheer expansiveness of the city. Banks of little white cumulus clouds pulse outwards towards the low green rise of the Southern Highlands. As I glance I realise that the suburb in the far southwest of the city where my participant lives, around 40 kilometres from where we are now, is probably just about discernible in the very far distance.

I begin by asking Kate about her commute into the city that morning. She throws her head back; her sharp, suited comportment creaking at the seams a little. It turns out that today she slept in by mistake and had to get a later train than usual. The discussion quickly moves on to talk about the different trains that she can take to work. She outlines how her choice is made by a number of different aspects. At one point she describes how it is the time of the journey that is important. She says "you've got two different types of trains: you've got a more express train and then you've got a slightly longer train that goes through the airport tunnel." Referring to the way that she uses the *TripView* mobile phone app to check the on time status, she implies that it is the express trains that are preferred: "So I'd just—as I'm getting ready, I'd just grab it here on my phone and just—and just check to see what might be an express train that's going to get me to work reasonably on time." But speed isn't everything. Kate goes onto say:

> The other thing is I'm always thinking, 'Is it going to be a good train that's air conditioned, or is it going to be one of the old rattlers?'—because, obviously, I prefer the more air-conditioned trains and that is definitely a factor in deciding which trains I catch each day.

Confirming the complexity of her decision, she emphasises "It's not always that the express trains are the good trains or anything; it's just—I just know that, yeah, when the good trains are going to come."

Framing this discussion through what has already been written on comfort in terms of the desirability of different technologies of transit, hearing from Kate that she prefers the air-conditioned newer trains over the older "rattlers" is perhaps not too surprising. However, the remainder of the interview complicates these assumptions about comfort in interesting ways.

In Kate's descriptions of today's morning commute, she reflects on the effects of travelling through the airport tunnel:

> So this morning it was coming via the airport tunnel, which is something that—I'm not really thrilled about coming through that tunnel, because you're in the light and then you're in the dark and it kind of puts you to sleep. And then you come back out at Central and you're blinking, like: "Oh, God".

Here, we get the sense that the bright dazzle of emerging from the tunnel at Central station generates discomfort. Soon after, when the discussion turns to activities during the commute, we get the sense that it is being asleep itself that is uncomfortable:

> Generally, I try and have a book with me and, if I—at the moment, I'm studying as well—I'm studying for my [program name]—and, when I run out of books, I then study. But I'm lonely on the train and often I will—I might read a couple of pages and not really absorb much and fall asleep. So that's—I go through phases of—yeah; I have a few days of just falling asleep on the train, which is not overly comfortable.

Re-reading the interview transcript, it is hard to get a sense whether it is the physicality of sleeping itself that is not overly comfortable, or whether it is her lack of productivity that generates discomfort, as the following observation seems to suggest:

> I see a lot of people on the train just constantly, like, sleeping. Like, people will bring jumpers and put them on the window and sleep like that; and I think, 'Yeah, but'—you know—'you could be utilising this time a little bit better.'

Kate's evaluation of what for her makes for a comfortable commute becomes more clearly articulated a bit later on in the interview when I ask her whether she ever drives to work. She says that she occasionally drives, but expecting her to flag its stressful nature because she doesn't do it frequently, I am surprised by her response. She says:

> Occasionally—like I—when I drive, I feel like I get to work a lot more alert and I have a lot easier day because I'm switched on straight away and I'm constantly thinking about what's going on on the road.

I reply, asking "So your attention is a lot sort of sharper?" She replies:

> Absolutely, yeah; compared to getting on the train and then just not really having to think, I think—I think you're awake and you're going

> and then you feel, 'Okay, yeah, this is good yeah.' You get on the train and I think it really pushes your energy levels back down and I think it does—I think it does affect me for the rest of the day.

When I ask her what she means by this she relates this directly to her work performance. She says "I don't generally feel as awake as if I have driven to a client...I've got to switch on fairly quickly when I get on work. And, yeah, I think the train just doesn't—doesn't help that."

In contrast to the commonly-flagged idea that it is the agitations of the commute that are uncomfortable and need to be reduced, Kate's reflections on her different commuting experiences in Sydney suggests that for her it is the absence of agitation that is uncomfortable in terms of her body not being prepared for the day ahead. I challenge her directly about how some people have talked about the negative effects of the agitations of driving to work. Confirming this position, she replies quite bluntly "I'm the opposite". She follows this by sharing with me some of the strategies that she has developed to deal with the sedative effects of commuting by train:

> If I get out of the train and I—like, I might have had a late night and I know I look pretty tired, I will maybe just—like, I just recently had a client just down at [place] and I'll get off at St James and I might just do a quick walk around the park and then go to work, just to try and wake myself up a bit, depending on if I've got my laptop and whether I'm carrying heavy things. But sometimes, yeah, if I know I've got time, I might just do a quick walk just to sort of wake myself up and become acquainted with outside, because I rarely see it in the mornings before—before I go into the office.

She goes on to suggest how:

> If I was on the bus, I'd jump off the bus a stop early and sort of walk the rest of the way, or I might—I generally do think of things—you know, ways I can go out of my way to get in a little bit more exercise around my travel. It is not so—not so prevalent with the trains, because it's all set; but, on the buses, it was a little bit easier; like, you might walk to the next stop when you get on and that sort of thing.

Some interpretations

The interview provided a space through which both participants could reflect on the comforts and discomforts of their commute. Revisiting Ahmed's (2004) contention that comfort suggests an "ease and an easiness" in terms of the way that bodies inhabit the world, what is particularly striking is how the two interviewees draw out quite different accounts of what constitutes a comfortable commute for them. Indeed, in the process of analysing each of my interview transcripts, it is this distinction that drew me to think about these two interviews, presenting them side by side in this chapter. Whilst undertaking a more "systematic" interpretation inevitably obscures some of the nuances and specificities of each interview, I want to spend the remainder of this chapter dwelling on this distinction because it will potentially help to problematise the concept of comfort further.

A central theme of Helen's interview was the challenge of how to move between home and work following an accident and subsequent illness that rearranged her bodily capacities. This involved cutting excess movements and slowing down; removing herself from exposure to strong smells; and moving in such a way that protects her right arm and hand. So Helen's commute involved removing potential agitations and reducing stimuli. Kate's interview on the other hand gives us a different idea of what makes a comfortable commute. A central theme of her interview was her concern for how the morning commute by train can leave her feeling sedated and not sufficiently ready for the work day ahead. In response, and in contrast to Helen, Kate's challenge was about finding ways to increase potential agitations and introducing stimuli to make her more alert.

Although this schematisation risks overplaying the distinction, there are some significant practical implications that can be drawn from this point of contrast. Perhaps most importantly, the first implication is that what makes for a comfortable commute is not the same for all commuters. This is because, as these two interview encounters illustrate well, commuters are different in terms of what their bodies can do and how the experience of the commute impacts on their bodies. Whilst this might sound intuitive, this variability contrasts with more essentialist assumptions about mobility hierarchies where some modes of travel are taken to be inevitably more comfortable than others. Relatedly, the second implication is that qualities of commuting that are often taken to be desired by all commuters such as relaxation, and the ability to get to work on autopilot for instance, are less universal than might be assumed. Whilst relaxation and "not having to think", to use Kate's words, might be desirable experiences for some commuters, for others these are experiences that are evaluated in a less favourable light, as Kate's reflections suggest.

These differences of how experiences of commuting are evaluated introduce some important conceptual questions about the relationship between comfort and commuting. On one hand, if we were to subscribe to an understanding of comfort that had some essential qualities, such as ease and an easiness, then the two interviews staged side by side would suggest that not all commuters desire comfort. On the other hand, we could take a more relational understanding of comfort where comfort is differently construed for different people. From this perspective it could be argued that Kate is more comfortable with being agitated on the commute than other people might be. Yet at the same time, both interview encounters relate to ease, albeit in different ways. Both Helen and Kate show a confidence about what is best for them in terms of getting through the day. For Helen, it is about making the commute gentle on her body to mitigate pain to ensure that she can make it through the day. For Kate, it is about having encounters during the commute where she is forced to think and react to set herself up for the remainder of the day. Both in a sense are about tactics to make life easier, particularly their capacity to meet their professional obligations.

Immanent evaluation of un/comfortable encounters: active and passive affections

I have previously suggested that comfort might be considered in three ways (Bissell 2008).[2] In the final part of this chapter, I want to take the third understanding of comfort as an affective resonance to explore this relational sense of comfort further and how we might understand it in terms of a process of learning to become active. In this respect, this discussion is informed by recent work in cultural geography guided by non-representational theories around the relational constitution of bodies through encounters. It provides us with a dynamic consideration of bodies that apprehends how they are individuated through the experiences that they have. Much of this thinking has been guided by Deleuze's rereading of Spinoza.

From this conceptual perspective, variations in a body's[3] power of acting are understood in terms of capacities to affect, or to actively give form, and capacities to be affected, or to passively receive form. A body's power is variable in this respect because it is immanent to the encounters that each body has. It is the distinction between increases in a body's power of action and reductions in a body's power of action that provides the basis for a real ethical difference (Deleuze 1990). What makes this conceptual frame so useful is that it gives priority to understanding how it is the ongoing encounters that people have, however big or small, that change that body's power to act. This relational understanding of power is different to a representational understanding of power where power is read-off from pre-existing, fixed identities. In terms of thinking about commuting, this turns the spotlights squarely on the encounters that make up the experiential texture of the commute itself. Commuting bodies are not fixed and inert, but are actually undergoing change through the encounters that they have. In this respect, bodies are "in a constant state of de-and re-composition in relation to other bodies, even in the most mundane acts of everyday repetition" (Ruddick 2010: 28). There is continuous variation not just in the sense that in the passage of events encounters succeed other encounters, but rather that each encounter is a variation in the power of acting.

Taking Spinoza's ethics as a point of orientation, transitions in powers of acting—affects—can be understood in terms of their activity and their passivity. Where active affections stem from bodies themselves, impressions onto bodies from the outside are passive affections. It is these passive affections that are particularly interesting from the perspective of commuting encounters. Spinoza makes an important distinction between sad passive affections and joyful passive affections. These can be understood in terms of two sorts of encounters. Deleuze's reading of Spinoza summarises these well. He says "the first sort occurs when I meet a body whose relation combines with my own... It produces in me an affection that is itself good, which itself agrees with my nature" (1990: 239). Joy is any impression involving an increase in a body's power of acting. A second kind of encounter happens when

> I meet a body whose relation cannot be combined with my own. The body does not agree with my nature, it is contrary to it, bad or harmful. It produces in me a passive affection which is itself bad or contrary to my nature.
>
> (1990: 241)

This impression corresponds to a reduction in a body's power of action. "We know what is bad only insofar as we perceive something to affect us with sadness" (1990: 241). Sadness is any passive affection that decreases a body's power of acting.

Returning to Helen's interview in this light and recalling her accident and subsequent illness, we can see that her body has a heightened susceptibility to passive affections that decrease her power of acting, for instance in the way that she describes how she might board a bus and smell perfume that does not agree with her body which diminishes her power of acting. Her power of acting is also diminished by the pain involved in being physically jostled around. Turning to Kate, on the other hand, being jostled around does not result in the same diminution in her powers of acting. Being jostled around in the context of having to think when she's driving to work, for instance, increases her powers of acting. On the other hand, the relaxed encounter with the train carriage is evaluated by her body in a way that decreases her powers of acting. An important aspect to draw out here in terms of understanding commuting comforts therefore involves *learning to distinguish* between joyful passions and sad passions: what feels good and comfortable and what feels bad and uncomfortable which can only be evaluated from the perspective of that body over time. As Deleuze points out, "nobody can undergo for us the slow learning of what agrees with our nature, the slow effort of discovering our joys" (1990: 262). This is the first part of the process of learning about comfort.

In terms of the desire to be comfortable Spinoza's relational ethics give us a sense of why we prefer to seek out comfortable encounters that accord with our nature. Yet as Ruddick (2010) points out, the desire to reproduce joyful encounters with bodies that accord with our nature and avoid painful ones with bodies that do not accord with our nature poses some important questions concerning encounters with difference that produce discomfort. Responding to this introduces a second aspect of learning about relations. Spinoza points out that all passive affections, even joyful ones, ultimately limit a body's capacity to act. Deleuze explains that this is because "our power of action is not yet increased to the point that we are active" (1990: 240). So even if the feelings of joy increase our power of action, "it never increases enough for us to come into its real possession" (2010: 241). If a body cannot grasp the cause of their affections, the way that they make sense of an encounter is inadequate, even for those encounters that feel good. The affection remains passive

> because it is explained by the external body, and the idea of the affection is a passion, a passive feeling. But it is a feeling of joy, since it is produced by the idea of an object that is good for me, or agrees with my nature.
>
> (1990: 239)

The important question then becomes: how can passive affections ever truly increase a body's capacity to act, if passive affections are ultimately taken to be limiting and restraining forces? In response, and following Spinoza's ethics, the most important ethical task is how to move from passive experiences of joy, to active experiences of joy. As Deleuze summarises,

> to do all we can amounts to two things: how exercise our capacity to be affected in such a way that our power of action increases? And how increase this power to the point where, finally, we produce active affections?
>
> (1990: 269)

This, for Deleuze, is the "art of organising encounters" (1990: 262) where our freedom is premised on extricating ourselves from chance encounters and sad passions to organise good encounters, combining our relations with other relations that combine directly with it. Significant here is that increasing our power *requires* us to have encounters with other bodies, human and non-human. In this sense Spinoza's ethics is a collaborative ethics. However, more than just having feel-good encounters, the second part of this learning process involves the formation of "common notions". This is about the adequate comprehension of causes. An example of this adequate comprehension of causes might be glimpsed from Helen's interview. Much of our conversation is characterised not by reactivity which might have been illustrated through angry statements about how much she might dislike commuting by bus, and thereby her power being reduced without comprehension of cause. Instead, our conversation is much more characterised by a detailed understanding of the bus's nature and speculating on the possibilities of becoming active with it to feel good. Helen's knowledge of the bus and of herself produces a "common notion" and a new relation, in other words, a new bus-passenger body that emerges from this specific context. Towards the end, her suggestion that "It's not that I hate the bus ride;

it's just the physical taxingness of it all. I quite like buses, you know" would seem to affirm the accomplishment of this activity.

Some conclusions

To paraphrase a key question raised by Sara Ahmed on comfort in relation to her writings on raced encounters is, quite simply: should comfort really be an aspiration? In *Queer Phenomenology*, rather than advocating for comfort, Ahmed reflects on the productive potentials of discomfort. In the context of experiences of discomfort when bodies fail to inhabit categories of identity, she says that

> discomfort... allows things to move. Every experience I have had of pleasure and excitement about a world opening up has begun with such ordinary feelings of discomfort; of not quite fitting a chair, of becoming unseated, of being left holding onto the ground
>
> (2006:154)

Ahmed's questioning of comfort when posed in terms of the cloying stasis of normativity relates to Ruddick's (2010) point about how a Spinozan ethics seems to caution against exposure to uncomfortable situations. This connection of discomfort to joy seems to contradict a Spinozist ethics where discomfort is a sad affection, in other words, a reduction in capacity to act.

Translating this suggestion back to the commuting sphere under investigation here would, on the face of it, seem quite unconvincing. Given the prevalence of commuter discomfort as a significant grievance in Sydney, making commuting *more* uncomfortable would hardly win the support of the city's commuting population. One could hardly imagine being disparaging of comfort, by suggesting that it is something that is negative. Yet the relations that generate experiences of comfort and discomfort in the sphere of commuting are complex. This comes into sharp focus when we think about the effects of some of the competing calls for changing commuting from different perspectives. For example, from an infrastructural perspective, some have argued that carrying out maintenance, repairs or upgrades to existing infrastructures is a disruptive process that has the potential to generate significant discomfort whilst work is happening (Graham and Thrift 2007). Here short-term pain becomes necessary for long-term gain. From a sustainability perspective, others have argued that discomfort will be an unavoidable aspect of adapting everyday practices to unexpected and perhaps volatile changes induced by climatic change (Gibson and Head 2013). Here, we will need to learn to live with more discomforts, for example through rationing of vehicle use for instance. Following the positive potentials of discomfort as Ruddick and Ahmed imply, in both these examples, discomfort has more positive connotations.

However, understanding comfort and discomfort from these aggregate perspectives, where comfort or discomfort is an inevitable outcome of a process or event, is less sensitive to the relational and variegated nature of comfort and discomfort as it is evaluated through ongoing experience. The divergent accounts of what counts as comfort for the two commuters in this chapter illustrates not only how aggregate accounts of comfort can be problematic, but that aggregate accounts also overlook the transitional way that events and encounters are registered by bodies and evaluated as being comfortable or uncomfortable, changing them in the process. In terms of commuting, there is an ongoing in-situ bodily evaluation of being in transit, following Spinoza's ethics, where we register that our power to act might be increased or diminished. But there are also spaces that can prompt reflection on the comforts and discomforts of our commuting practices. The interview encounters presented here might be one such space, replete with its own comforts and discomforts that envelop the recollections under discussion. It is these multiple sites of encounter and the passive affections that they are composed of that are instrumental in teaching about who we are, affording us opportunities for understanding and reflecting on the nature of our relational constitution such that we can actively create more joy and suffer less sadness in even the most everyday routine.

Acknowledgements

Thank you to participants at the 5th Aotearoa/New Zealand Mobilities Symposium held in Wellington in July 2014 for useful feedback on the ideas in this chapter. Thank you to Deborah Rogers for preparing the interview transcriptions. I am grateful for the support provided by the Australian Research Council Discovery Early Career Award (DE120102279) that enabled me to undertake this research project.

Notes

1 Interviewee names and places have been anonymised.
2 In this article, I suggested that there might be three ways that we could define comfort. First, we could think of it as an "objective capacity"; in other words, as an *attribute* of an object. Second, we could think of it as an "aesthetic sensibility"; in other words, as a distinctive bodily *sensation* that is produced by a particular environment. Third, we could think of it as an "affective resonance". I like this third definition because it invites us to think about how comfort is produced through the *encounter* of a body in an environment in a way that is much less determinate, as this section will show.
3 The term "mode" is used rather than body because Deleuze does not want to prioritise just the sense of a fleshy, human body.

References

Ahmed, S. (2004) *The Cultural Politics of Emotion*. London: Routledge.
Ahmed, S. (2006) *Queer Phenomenology*. Durham: Duke University Press.
Ahmed, S. (2012) *On Being Included: Racism and Diversity in Institutional Life*. Durham: Duke University Press.

Anderson, B. and Harrison, P. (2010) *Taking Place: Non-representational Theories and Geography*. Aldershot: Ashgate.

Bissell, D. (2008) 'Comfortable bodies: Sedentary affects', *Environment and Planning A*, 40(7), 1697–1712.

Dant, T. (2014) 'Drivers and passengers', in P. Adey, D. Bissell, K. Hannam, P. Merriman and M. Sheller (eds.) *The Routledge Handbook of Mobilities*. London: Routledge, 367–376.

Deleuze, G. (1990) *Expressionism in Philosophy: Spinoza*. New York: Zone Books.

Gibson, C. and Head, L. (2013) 'Anticipating forced change and extremity', Conference Paper presented at *RGS-IBG Annual Conference*, London, August 2013.

Graham, S. (2002) 'FlowCity: Networked mobilities and the contemporary metropolis', *Journal of Urban Technology*, 9(1), 1–20.

Graham, S. and Thrift, N. (2007) 'Out of order: Understanding maintenance and repair', *Theory, Culture and Society*, 24(3), 1–25.

IBM (2011) 'Australian cities not keeping up with commuter needs', IBM Commuter Pain Study http://www-03.ibm.com/press/au/en/pressrelease/33560.wss (accessed 10 January 2014).

Jain, J. (2011) 'The classy coach commute', *Journal of Transport Geography*, 19(5), 1017–1022.

Richardson, T. and Jensen, O. (2008) 'How mobility systems produce inequality: Making mobile subject types on the Bangkok sky train', *Built Environment* 34(2), 218–231.

Ruddick, S. (2010) 'The politics of affect: Spinoza in the work of Negri and Deleuze', 27(4), 21–45.

Saulwick, J. (2011) 'Drive to work like 15 rounds with Muhammad Ali: Auditor-General', drive.com.au online at http://www.drive.com.au/motor-news/drive-to-work-like-15-rounds-with-muhammad-ali-auditorgeneral-20111130-1o5ww.html (accessed 27 May 2014).

Stewart, K. (2013) 'Regionality', *Geographical Review*, 103(2), 275–284.

3 Beyond the 'comfort zone'

Experiencing and responding to everyday weather

Eliza De Vet

Introduction

Climates are changing, temperatures warming, weather patterns varying and extreme events intensifying. Yet, rather than facing change, industrialised societies are spending 90 per cent of their time indoors (Kosonen and Tan 2004), increasingly in air-conditioned buildings. This indoor retreat is problematic for two reasons. First, the energy required to maintain static indoor environments is resource intensive, contributing to greater climate change (Roaf *et al.* 2010). Second, concerns have been raised over potential air-conditioning 'addiction' (Brager and de Dear 2003; Cândido *et al.* 2010). Through exposure to thermally monotonous conditions, societies' capacity to tolerate change maybe diminishing, as vernacular strategies are no longer necessary (Hitchings 2010). In a spiral of increasing dependence, air-conditioning is exacerbating the effects of climate change, while potentially relegating societies' capacity to withstand and respond to changes in weather and climate.

To intercept, even curb these trends, engineering and sociocultural research has begun identifying broader notions of comfort and resource-efficient comfort strategies. This chapter discusses these pursuits, before examining comfort more broadly. As the majority of thermal comfort research has justifiably sought solutions from inside buildings, this chapter takes a step outwards, analysing everyday weather experiences and responses. Specifically, it explores how individuals find comfort under dynamic weather conditions in a variety of spaces (both indoors and out), involving different daily activities. This chapter examines accounts from two Australian study areas with contrasting climatic and cultural contexts – temperate Melbourne and tropical Darwin. These observations are drawn from a larger research project investigating the role of weather in everyday life (see de Vet 2014). Insights highlight the complex and diverse nature of comfort and the existence of effective vernacular response strategies. Through these accounts, attention is drawn to the problematic use of 'comfort' in endeavours to reduce the energy demands of air-conditioning.

In the pursuit of resource (in)efficient concepts of thermal comfort

Since the 1920s, when indoor spaces could not only be heated but cooled, enhancing comfort has been a preoccupation of engineers. As the efficiency and accuracy of heating and cooling devices improved, standards and expectations have constricted (Meyer 2000; Healy 2008). International standards and guidelines for air-conditioning vary but include those set by the American Society of Heating, Refrigeration and Air-Conditioning Engineers (ASHRAE), the International Standards Organisation and the European Committee for Standardisation. The global proliferation of air-conditioning, in both commercial and domestic buildings, has been driven by expectations for physical comfort, economic productivity and western fashions (Sahakian and Steinberger 2011; Nicol *et al.* 2012). In most commercial buildings, thermal settings are the result of 'heat-balance' or 'static' models, derived from chamber experiments. Calculations take into account metabolic rate, clothing insulation, air and radiant temperatures, air speed and humidity (ASHRAE 2013). These models have been considered universally applicable regardless of local climate and culture (Roaf *et al.* 2010). While air-conditioning standards differ, temperatures between 21°C and 25°C are generally recommended in order to achieve an 80 per cent occupant acceptability rate (ASHRAE 2013)[1].

Maintaining such narrow indoor temperatures is energy intensive (Cândido *et al.* 2010), particularly where there is a considerable temperature discrepancy between the indoors and out. In 2002, 33 per cent of global greenhouse gas emissions were produced by the building sector (Levermore 2008), substantially inflated by the energy demands of air-conditioners. In 2006, the American building sector accounted for 39 per cent of national energy demands, 35 per cent of which are attributed to air-conditioning (Kwok and Rajkovich 2010). Across the globe, air-conditioner dependence is on the rise, and Australia is no exception. Between 2005 and 2011, domestic air-conditioner ownership rose from 59 per cent to 73 per cent (Australian Bureau of Statistics 2011a). By 2020, this rate is expected to skyrocket five-fold compared with 1999 levels (Strengers 2010). In Australia and elsewhere, this rapid growth is causing blackouts during hot days and heatwaves as energy infrastructure fails to meet demand (Hajat *et al.* 2010; Maller and Strengers 2011). In the absence of energy-independent coping strategies, blackouts pose a lethal situation. The seriousness of blackouts is intensified in air-conditioner dependent buildings due to a general lack of passive heating and cooling designs (such as operatable windows; Pearlmutter 2007; Roaf *et al.* 2010). To combat accumulating air-conditioner complications, research is pursuing avenues to further expand comfort understandings and practices.

While engineers were central to the constriction of comfort standards, they have also been instrumental in efforts to re-stretch comfort concepts.

Fundamental to these efforts has been the introduction of alternate 'adaptive' comfort models. Adaptive models acknowledge the ability of building occupants to keep comfortable using a range of behavioural, environmental and technical adjustments. Currently, these models only apply to naturally ventilated buildings (Nicol *et al.* 2012). Adaptive models, which utilise 'field' observations, have identified broader comfort zones. For example, in Nicol *et al.*'s (1999: 277) commonly cited Pakistani study, involving 34 buildings within five different climatic regions, results indicated 'there is generally little discomfort at indoor globe temperatures between 20 and 30°C'. A host of other studies have published similar findings, pushing comfort boundaries as high as 33.1°C (Indraganti and Rao 2010). Adaptive models are reducing environmental and economic running costs of buildings. However, while 'comfort zones' are broader, air-conditioning still offers a solution to avoiding zones of discomfort.

To push boundaries further, social and cultural research is contesting the assumption that air-conditioners are indispensable. In 2003, Elizabeth Shove ignited discussions on the social, cultural and technical influences involved in the (re)production of current unsustainable comfort concepts and practices. Ensuing studies have worked to deconstruct current comfort understandings and contemplate potential avenues to redirect interpretations. For example, Strengers (2008) suggests unsustainable peak energy in Australia is the outcome of consumer demands, as well as government, industry and utility providers' perception of Summer – a time seen as synonymous with air-conditioning – to escape rather than confront outdoor weather. These perceptions are transferred to customers through energy pricing and the promotion of cooling devices. As such, Strengers (2012) argues that consumer attitudes, behaviours and choices should not be seen as autonomous and solely responsible for demand. Studies have also documented diversity among alternate strategies for comfort and underlined factors that limit or facilitate their implementation. For example, Fuller and Bulkeley (2013) discovered that deep-seated cultural expectations and norms prevented a small number of English migrants from embracing Spanish cooling strategies in the warmer climate. Gram-Hanssen (2008) reported that perceptions of a 'good life', including the importance placed on relaxation, physical activeness and outdoor interaction, was responsible for differing household heating practices. By unpacking current comfort concepts and practices, sociocultural research is working to reconfigure and direct comfort understandings and standards towards more resource-efficient futures.

Research on the ability of individuals to cope with temperature variations in the absence of air-conditioning is gaining traction. But there is more work to be done. Research needs to comprehend the complexities of comfort in everyday life and not just within buildings. It is essential research continues to extend into other spaces, particularly those outdoors, in order to discover the various ways comfort is conceptualised and achieved. As Hitchings (2009: 91) suggests, 'we might want to take some steps away from the "building" and begin with the places where different groups are still inclined to cope with a variety of immediate environmental conditions'. Equally important, is the need to examine different climatic and weather conditions. As most research has focused on temperate locations, investigation into other climates is warranted. Further, research must begin to comprehend current weather-relations and not simply temperature (see de Vet 2017). It is true, temperatures are rising, but these changes are part of broader weather variations. By exploring comfort in various spaces, under different weather and climatic conditions, research is better placed to identify resource-efficient approaches to indoor comfort. To begin addressing these research requirements, the remainder of this chapter explores everyday weather experiences and responses in temperate and tropical areas of Australia.

Studying Australian weather

Australia's climate is diverse, ranging from equatorial to temperate (Figure 3.1). To begin understanding how different Australian climates are experienced and responded to, two study areas were selected – Darwin in the north and Melbourne in the south (see Figure 3.1).

Established around the Yarra River in 1835, Melbourne (Victoria) is home to a growing population of four million (Australian Bureau of Statistics

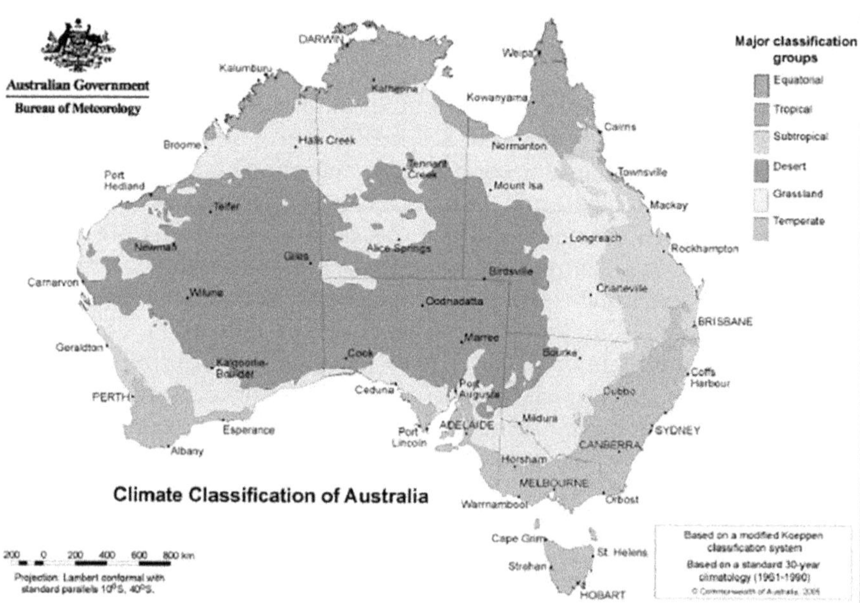

Figure 3.1 Study area location and climate classifications based on Köppen's classification system.

[ABS] 2011b). Melbourne is Australia's second largest capital city (after Sydney), hosting a large business hub and attracting significant tourist numbers. Its residents experience the four traditional European seasons – Summer, Autumn, Winter and Spring. According to Köppen's classification system, Melbourne has a temperate climate, having warm Summers and no dry season (Bureau of Meteorology [BoM] 2008). During Summer, hot days and heatwaves are frequent, with temperatures reaching as high as 46.4°C. Winters are moderately cold, with daily July temperatures ranging between 6.0°C and 13.5°C (BoM 2014a). Compared with northern cities, Melbourne has reliable but lower rainfall with fewer thunderstorms and severe storms. Legendary for its 'four seasons in one day', its changeable weather is the outcome of its coastal position, where warm continental temperatures meet cold Bass Strait air. These weather changes warrant visitor advice – 'A tip for any visitor is to be prepared for anything – take an umbrella and wear layers that can be worn or removed as needed!' (City of Melbourne 2011: n.p.).

In the north of Australia, Darwin's climate is categorised as 'tropical savannah', having two seasons – 'the Wet' and 'the Dry' (BoM 2008). 'Tropical savannah' reflects its warm and relatively stable temperatures, where average daily maximums range from 33.3°C in November to 30.6°C in June and July (BoM 2014b). Unlike its temperatures, Darwin experiences substantial variation in monthly rainfall – over 425 millilitres in January and 1.2 millilitres in July (BoM 2014b). In addition to the Wet and Dry seasons, local indigenous and non-indigenous residents acknowledge a third distinct season – 'the Build-Up', a period of high humidity. In this chapter, all three seasons are recognised – the Build-up, the Wet, and the Dry. Darwin's tropical location also exposes it to cyclonic events. Since its European colonisation in 1869, the city has been demolished by cyclones on three occasions – 1897, 1937 and 1974. The most recent and devastating, Cyclone Tracy, changed many aspects of Darwin life, including its architecture, which saw transitions from passive tropical designs to air-conditioner dependent styles (Parish 2007; Rothwell 2007).

To investigate the role of weather in everyday life, a longitudinal study was conducted involving 20 Melbourne and 16 Darwin participants. The study areas were selected to highlight diversity in experiences of comfort. Accounts were generated using two approaches, each applied to one of the two study areas (for details, see de Vet 2013). While approaches differed, they focused on participants' weather experiences and responses while occupying different spaces and engaging in a range of activities. Discussions varied from household chores, commuting, and leisure activities, to naturally ventilated home and workspaces. How participants felt physically during the course of the day, including sensations of comfort and discomfort, were central to questions. The following two sections draw from one of the two study areas – thermal comfort in Melbourne and comfortable weather conditions in Darwin.

'I enjoy heatwaves': identifying and understanding diversity in thermal comfort in Melbourne

> I like it hot. Like, really hot. I'm quite happy for it to be 40 degrees or something. I enjoy heatwaves.
>
> (Levi, Spring)

To find heatwaves comfortable is perhaps surprising. However, Levi's statement suggests that perceptions of comfort may involve more than physical sensations. In Melbourne, participants' thermal comfort diverged by 26°C, extending between 14°C and 40°C, with most preferring temperatures of 25°C and above. Other weather elements were also important, notably humidity, but temperature was always influential. Beyond these levels of comfort, however, was not 'discomfort'. Rather, participants identified temperatures that were neither comfortable nor uncomfortable, encapsulated both simultaneously, or included brief but acceptable periods of discomfort. These mid-point positions were often described as 'okay', 'not perfect', 'sufficient', 'bearable', or 'endurable'. Participants expected and accepted these 'manageable' conditions as part of everyday life.

Differences between engineering definitions of comfort and participants' thermal experiences are likely the outcome of discrepancies in methodological approach, namely: subjective responses; variation in terminology (i.e. 'comfortable', 'preferred', 'optimal' and 'ideal'); the 'everyday' research focus, incorporating a variety of activities with differing levels of physical exertion; and, the inclusion of outdoor and transitional spaces. However, in examining participants' justifications, intriguing understandings of comfort came to light.

Health and physiology

Health issues and physiological cycles and transitions affected participants' thermal comfort, either permanently or intermittently. For two participants, painful chronic illness symptoms, exacerbated by the cold, fed into preferences for warmer temperatures. Sleep deprivation also elevated desires for warmer temperatures, but only ephemerally. Conversely, preferences for cooler temperatures correlated with changes in oestrogen levels. One participant reported overheating at night when menstruating, no longer needing blankets in bed. Another participant, Miranda, found sticking her head in the freezer or ambling in refrigerated supermarket isles provided relief from menopausal symptoms. For Miranda, not only could her preferred temperature be impacted immediately by hot flushes, but over the past few years her preferred temperatures permanently changed from 25–28°C to 20–21°C.

In addition to the health-related issues above, simple differences in comfort were observed between participants and their partners. These differences were identified when interviewing couples or when participants made comparisons with their partners. These differences were interesting as couples frequently engaged in the same activities. Differences were exemplified when discussing levels of clothing and bedding, and desires to change the thermostat. For example, while Margret slept under bed sheets, her partner went without, and turned the air-conditioner down – 'my husband's – oh, God. If he could sleep in Antarctica, he would, bare bum.' These health and physiological matters suggest that while physiology is central to calculating universal thermal comfort, variations commonly affect comfort levels. Moreover, variations not only occur between participants, but participants' own comfort level.

Residential history

Having previously lived in other climates – New Zealand, England, Fiji, Hong Kong, Queensland, or regional Australia – eight participants used their residential history to justify their thermal preferences. As most of these participants had resided in Melbourne for years, even decades, physiology could not be used to explain the role of residential history in comfort. Health research suggests that physiological acclimatisation occurs over days to weeks as the body adjusts to new conditions (although long-term adaptations are still being discussed; Stinson *et al.* 2012). On closer inspection of participants' accounts, residential history appeared to influence preferences through nostalgia, as particular temperatures reminded participants of past places, 'home' and their younger years.

The effects of residential history also impacted comfortable temperatures in other ways. For instance, Amelia had previously worked in a hot rural region of Australia. But, rather than increasing her tolerance for warmer temperatures, Amelia's ideal temperature was what 'you're supposed to set the air-conditioner at … between 20 and 22°C'. She had retained her past behavioural adaptations, which affected her perceptions of comfort. While Amelia's account demonstrates signs of air-conditioning addiction, she found her naturally ventilated home and workspace satisfactory and made adjustments as required. Whether residential history influenced participants' thermal perception through behavioural responses, past memories or other avenues, it was clear it had a notable impact on participants' comfort.

Activity association

Comfort also reflected temperatures that afforded enjoyable activities. Levi, who enjoyed 40°C and heatwaves, explains:

> I will cover myself in sunscreen and a hat, but I'll be wearing shorts and a Hawaiian shirt or something, and I guess I'll just wander around. I mean, it sucks if you have to do anything, but usually, during that time,

> it's Christmas. So you don't have to do anything. It's summer holidays. So I'll usually wander around or be drinking beers in the backyard or reading a book or something.

For other participants, preferences for warmer temperatures were associated with frequent physical and social activities, such as family gatherings, nights out, and trips to the beach. These activities were not only connected to warmer temperatures but also to longer days. No longer returning home from work in the dark, warmer and lighter times of the year encouraged after-work activities – evening walks with family, playing outside with children and dogs, pottering in the garden, dabbling in photography and meeting up with friends. These periods of heightened activity were a stark contrast to the cold and dark periods of the year, which generated negative moods and emotions and zapped energy levels – 'you tend to hunker down in the winter. Sort of hibernate… you just don't feel like going out' (David).

Activity association also worked at a micro-level. For a number of participants, warmer weather was associated with one layer of clothing, which they felt was less constricting and simple to coordinate and clean. But warmer and lighter clothing was not always appealing. For Michael, his sense of security and confidence in social situations was tied to clothing layers – 'Clothes are like armour … I am protected.' The desire for more layers contributed to his preference for colder conditions. Whether activity associations were simple or more substantial, enjoyable activities also influenced participants' level of thermal comfort.

Health and physiology, residential history and activity association were all implicated in Melbourne participants' broad perceptions of thermal comfort. Interestingly, health and physiological issues are not accounted for in ASHRAE (2013: 4) standards, as 'the body of available data does not contain significant information regarding the comfort requirements of children, the disabled, or the infirm'. By excluding individuals with health issues or those undergoing physiological change, comfort diversity within and between individuals is obscured. Participants' experiences also demonstrated that comfort is not only a product of sensation but residential history and activity association. These findings reflect current engineering discussions, which utilise psychological factors to explain comfort differences, particularly in naturally ventilated buildings and outdoor settings (see for example Nikolopoulou and Steemers 2003). Brager and de Dear (1998: 87) suggest:

> psychological adaptation encompasses the effects of cognitive and cultural variables and describes the extent to which habituation and expectation alter one's perception and reaction to sensory information … [where] optimal adaptation levels are established as functions of past exposure and act as benchmarks or norms for environmental evaluation.

Accounts from Melbourne participants support assertions that thermal comfort exists well beyond engineering definitions for air-conditioned buildings. By advancing comfort calculations beyond physical sensations to include health issues, physiological cycles and transitions, and psychological influences, another opportunity to loosen comfort standards is presented. While drawing comparisons between everyday contexts and air-conditioned spaces may be considered contentious, disputes only accentuate narrowly fashioned expectations for comfort and strict purposes inscribed into air-conditioned buildings.

'The sea breeze makes it very pleasant': comfortable Darwin weather conditions

Most Darwin participants shared similar ideas of comfortable weather – mild temperatures (around 28°C), with low humidity and a breeze. However, different combinations of weather elements could also provide comfort or ease less-than-comfortable conditions. Similarities between participants' preferences were likely the outcome of relatively consistent and predictable weather conditions that frequently presented physiological and psychological challenges.

Temperature

Darwin participants had an affinity for temperatures well above 25°C, mirroring findings for hot and humid regions (see for example Lin 2009; Indraganti and Rao 2010). A cold-snap towards the beginning of the Wet research week, reducing daily maximums from the low 30s to low 20s, was noticeable in an otherwise consistent climate. During this time, the temperatures that got 'right down to 22' (Kathleen) were described as 'quite cool', 'chilly' and 'cold', requiring participants to 'brave the conditions' (Eddy), giving the 'feeling as though I could curl up and go to sleep' (Trudy). Similar comments were made by some participants during a particularly cold Dry season, when daily maximums reached 28.2°C, with minimums of 13.5°C. During the colder parts of the day, participants commented on the 'freezing cold' (Indre) and 'chilly mornings … [and the] winter freeze' (Lucy), where one had to 'dress for the worst possible scenario … [as] it just wasn't warm enough for my liking' (Jenny).

Conversely, 'warm' and 'hot' weather was common throughout the year, with daily maximums wavering between 30°C to 34°C. These familiar and expected conditions were an accepted part of Darwin life. While maximums were considered 'hot', conditions were frequently tolerable, even enjoyable, depending on activities conducted at the time – 'What was the weather like today? Hot but ok, most of day spent indoors, under verandas or in the car' (Bernadette). Only one participant, Geoff, found these temperatures uncomfortable, no matter the activity. His reluctance to adopt climate-appropriate adjustments potentially contributed to his dislike for warmer conditions (see de Vet Forthcoming). After a short six-month stay in Darwin, Geoff's preference for temperatures 27°C and below, influenced his decision to move back to Australia's temperate south.

As daily maximum temperatures rarely exceeded 34°C, conditions described as 'too hot' or 'very hot' referred to temperature duration rather than degree. Uncomfortably hot temperatures resulted from an absence of relief, including anticipated cooler nights – 'It is a very hot night. My husband looks at the thermometer and it reads 34 inside' (Laura). The majority of participants were tolerant of warm and hot temperatures, so long as cool changes were regularly experienced, providing physical and mental rejuvenation:

> While hot and sweaty, days like today are not unusual for fieldwork in the Build-Up, the weather makes it a great time of the year and I love being out in it all. You know you are alive when the thunder is crashing, lightning flashing and the cooling gust of wind at the front of a storm cell hits you in the face.
>
> (Ben, Build-Up)

As suggested from Ben's extract above, comfortable weather involved more than temperature.

Humidity

In Darwin, humidity was often inseparable from comfort and important to perceptions of heat, particularly conditions that were 'too hot'. High humidity during the Build-Up and Wet was found to compound hot uncomfortable temperatures, often flustering and draining participants more than the heat:

> Sometimes I notice changes in humidity, but am always hot. Humidity sometimes makes me feel less energetic and it is easier to feel frustrated.
>
> (Anne, Build-Up)

During the Dry, participants celebrated low humidity and daily maximum temperatures, conditions that compensated for the previous less-than-comfortable seasons. However, not all participants enjoyed low humidity and temperatures. For Indre, a participant originally from Sydney, the Dry weather was too much:

> The weather this time of year is definitely affecting my lifestyle. I feel sluggish and depressed, … I'm watching more DVDs than normal, eating more than normal, exercising less. I think this is all heightened by the fact that a lot of people in Darwin are thrilled with the weather at the moment – and I just want the humidity to come back!

Wind, rain and cloud cover

Uncomfortable combinations of heat and humidity were alleviated by wind, rain and/or cloud cover. A breeze that moved stagnant hot and humid air was welcomed, while rain and cloud cover, so often negatively perceived in other urban areas of Australia, were yearned for and celebrated during the Build-Up:

> 18:30 We cycle to the beach to catch the sunset, very little cloud in the sky. The sea breeze makes it very pleasant, just a comfortable temperature.
>
> (Sarah, Build-Up)

> Finally, relief from the heat! … It's not a big storm (yet!) just a light shower, but I can already feel the life being breathed back into the dry landscape … My body feels alive and refreshed again.
>
> (Jenny, Build-Up)

> It was pleasant without the sun out.
>
> (Laura, Build-Up)

But the relieving capacity of these elements had limits. During the Dry, when temperatures and humidity were at their lowest, rain, wind and cloud cover could make for uncomfortable 'wintery' sensations. Warming sunshine, once avoided, was now sought after in the cooler morning and afternoons. In addition, comfortable rainy conditions, like temperature, were dependent on duration. During the Build-Up, participants eagerly anticipated rain to relieve high humidity and temperatures. As rain became expectantly more frequent, participants became impartial to wet conditions. Towards the end of the Wet, after weeks of rain, some flooding, and lowered levels of heat and humidity, rain became repressive, causing negative moods and emotions and a sense of becoming 'cabin feverish' (Jane). As these accounts suggest, different weather elements could provide both comfort and discomfort.

By looking outdoors, away from the shelter of air-conditioned buildings, experiences of weather serve as a reminder that comfort is not only a condition of temperature. Like temperature, humidity, wind, rain, cloud cover and sunshine, provide both comfortable pleasures and sensory challenges. Through these daily encounters, sensory experiences are broadened and the range of 'acceptable' conditions is expanded. Maintaining the capacity to abide weather and its variations will be essential to coping with greater future environmental change.

'Work while it's cool': vernacular weather adjustment strategies

Participants in both Melbourne and Darwin adjusted to weather on a daily basis. Adjustments included both energy-intensive and non-intensive strategies.

Most participants used energy-intensive heating and/or cooling devices at some point throughout the year. During cooler Melbourne weather, all participants used heaters while at home, work and in the car. Over the warmer months at home, they used electric fans, while air-conditioners (owned by two-thirds of participants) were reserved for hot summer afternoons and evenings. In Darwin, most participants were exposed to air-conditioning at work. But at home, exposure was a different story. Five participants did not own or never used an air-conditioner, and all other participants restricted air-conditioner use to the Build-Up. While participants in both study areas were exposed to heating and/or cooling at some point, they recognised the benefits of non-energy-intensive weather responses to remain comfortable. The following section examines key non-energy-intensive strategies and the features that facilitated or limited these responses.

Melbourne

Of all Melbourne weather response strategies, personal adjustments were most common. Greatest diversity occurred in the home, afforded by privacy and control. Away from the public eye, participants wore very little or donned bulky, unflattering clothing, even wearing dressing gowns and outdoor jackets over day clothes. Participants sat under blankets, moderated physical activities, adjusted the type of food and drinks consumed, moved to cooler or warmer parts of the home, and invented new approaches to enhance comfort (such as hanging blankets from upper bunk-beds to trap heat while sleeping). The quality and quantity of bedding was particularly important to all participants (even worth stealing from other family members), as it avoided the need to use heaters overnight. These strategies contribute to a growing list of personal adjustments that have been documented globally (see for example, Strengers and Maller 2011; Fuller and Bulkeley 2013).

Adjusting the surrounding environment and altering fan settings were also effective weather responses. By simply opening or closing doors, windows, blinds, curtains and adjusting fan settings, changes in temperature and air movement often made all the difference. The effectiveness of these strategies has been noted in other studies (for example see Hwang *et al.* 2009; Haldi and Robinson 2010). While adjustments were facilitated by relaxed social expectations in the home, they were also encouraged by financial and environmental incentives, and the sense of accomplishment experienced in achieving comfort independent of heating devices.

Outside the home, personal adjustments were important as highly structured daily routines left participants little room to coordinate activities around the weather. At work, while commuting, and during scheduled social and leisure activities (for example, family gatherings and team sports), adjusting clothing, using sun and rain protection and consuming warm or cold food and drinks, were essential to remain as comfortable as possible. However, participants' ability to adjust was commonly hindered by poor infrastructure and ridged social expectations. In naturally ventilated workplaces, these included a lack of fans, blinds and shade cloths, as well as stringent dress codes and work timetables. While commuting, work attire increased heat sensitivity, jammed train and tram windows prevented relieving airflow and expectations for arriving sweat-free at events increased the need for car air-conditioning. At home, leaving doors and windows open for cooler air was not always an option due to windows that were painted shut, security fears, and building designs that created hazardous wind tunnels and caught street noise. In these instances, participants' control over their environment was restricted. Reflecting a wealth of research on adaptive comfort and personal control (Nicol *et al.* 2012), these limitations reduced participants' tolerance for thermal variation, resulting in or amplifying uncomfortable sensations. If these obstacles to energy-efficient cooling strategies reflect broader Melbourne trends, a significant opportunity to intercept the rise of air-conditioned buildings is apparent. Rectifying these barriers will not only enhance adjustment effectiveness and opportunities, but increase individuals' sense of control over their environment and tolerance for thermal variation. Enabling adjustments will be crucial to normalising everyday resource-efficient adjustments and avoiding demands for air-conditioning.

Conversely, during free time when schedules were flexible, participants responded to weather differently. As leisure time was to be enjoyed and not wasted and weather held the potential to enhance or ruin experiences, weather weighed heavily in activity decisions. Participants took advantage of particular weather conditions, and as such activity frequency changed throughout the year. Social gatherings, exercise sessions and outdoor activities all increased with warmer temperatures and longer days. Conversely, these activities were often avoided, cancelled, moved inside or cut short with cold temperatures, rain, cloud cover, extreme heat, high humidity, excessive wind, ultraviolet-radiation and/or during the shorter days of the year. These findings reflect similar observations in research on physical activities (see reviews by Tucker and Gilliland 2007; Chan and Ryan 2009).

Given the opportunity, Melbourne participants showed substantial ability and willingness to respond to the weather using resource-efficient responses. However, waiting for perfect conditions before going outside presents ramifications for long-term weather connectivity and wellbeing (Hitchings 2010, 2011). For this reason, it is important to find and support strategies that coax individuals outdoors, particularly if weather under climate change further discourages them from stepping outside.

Darwin

Coping with Darwin weather without air-conditioning was important to all participants (with the exception of Geoff, discussed above). Participants recognised that prolonged exposure to air-conditioning lessened their ability to deal with outdoor conditions, while inflating financial and environmental costs. Such observations have not changed since Williamson *et al.*'s (1989) earlier Darwin study. To cope with Darwin's tropical climate, particularly during the Build-Up, participants utilised a range of adjustment strategies. These included: wearing sun protection and light clothing made of natural fabrics, drinking plenty of cool fluids, consuming light and cold foods, resting during the heat of the day or moderating daily tasks, taking cool showers, and actively seeking shade and a breeze (similar findings for warm-humid regions are noted in studies such as Indraganti 2010). Among strategies, two were particularly significant – climate-appropriate domestic property designs and strategic activity scheduling.

Climate appropriate domestic property designs

Climate appropriate houses and gardens, both long-term climatic responses, provided a robust foundation for keeping cool. Important external features included heavily vegetated gardens and verandas. All participants had gardens bordering their property and/or abutting the house, as they worked to 'cool the house and surrounds' (Kathleen). The following extract illustrates the thought and dedication participants gave to their garden and the rewards they received:

> These trees have mostly been planted by me and form an integral part of the shade and shelter that makes for a great place to live in a hot tropical climate. Strategically established a little away from the house on this northern side to allow air flow, while providing shade and greenery in the garden.
>
> (Ben, Build-Up)

In addition to gardens, nine participants had one or more covered verandas (up to four), which encouraged outdoor living and increased cooling shade around the property. Verandas offered spaces exposed to a breeze and sheltered from the rain and sun (enhanced by surrounding vegetation). They were highly functional spaces, used for sitting, reading, eating and drinking, socialising, entertaining, cooking and more. Cooper (2008: 379) suggests in her reflections of the 1950s Californian garden and its cooling effects, that 'the best designs for functional gardens incorporate spaces for both work and leisure, mirroring the diverse activities within the household'. In Darwin, these spaces encouraged outdoor living by blending the indoors with the out, as participants introduced fans, lights, dining sets, lounges and even refrigerators.

When outdoor conditions became too warm, stagnant and sticky, participants retreated to more comfortable indoor spaces. Indoors, fans were crucial, generating conditions that were 'pleasant', 'relaxed', and 'comfortable'. Along with fans, each house had a unique range of climate-appropriate

Figure 3.2 Guy's photograph illustrates the climate-appropriate features of his property.

features. For example, Guy's house 'designed for living in the tropics' (Figure 3.2) incorporated high ceilings, polished timber floors (as opposed to carpet), open-plan living spaces, large banks of louvred windows with fly-screens, wide glass doorways (for airflow and natural light), blinds, multiple ceiling fans and a wide verandah with roof and retractable vertical shade cloths (reducing direct sunlight inside the house). Such climate-appropriate architecture was perceived as an asset that enhanced tropical living.

Strategic scheduling

Despite the tropical climate, participants routinely undertook physically demanding and outdoor activities, including gardening, exercise and commuting by foot or bicycle. To decrease encounters with discomfort, participants organised tasks around cooler times of the day. This involved scheduling activities for the morning and afternoon when temperatures were lowest and taking advantage of less predictable relieving weather conditions.

As the garden was an important resource for keeping the property cool, all participants gardened regularly. Scheduling was important and morning sessions began as early as 6 am, enabling participants to 'work while it's cool' (Cameron), 'before it gets too hot' (Trudy). By 9 am, the majority of gardening sessions ceased, recommencing just before 4 pm and continuing until dark. But escaping heat was just as important as avoiding direct sun. Most participants acknowledge the value of shade provided by overcast skies and trees:

> After 5 o'clock it was cool enough ... the heat of the day had gone and most of my garden at this time is in shade so it is much more tolerable.
>
> (Jenny, Build-Up)

While schedules appear to coincide with non-business hours, nearly half of all gardening sessions took place on the weekend.

Like gardening, exercise sessions were scheduled when temperatures and the sun were at their lowest, commonly before dawn and after dusk. However, unlike relatively predictable diurnal temperature and sunlight patterns, participants also seized cooler opportunities provided by less predictable patterns in wind, rain, cloud cover and humidity:

> Finally, relief from the heat! Rain started falling here 15 minutes ago and the temperature outside is now much closer to the natural temperature inside ... Well the rain has stopped now, so I'm going to take that walk now.
>
> (Jenny, Build-Up)

Participants' ability to coordinate their activities around the weather was important to avoid long periods of discomfort. Part-time or flexible work hours and activities enabled participants to adjust accordingly. Equally important was participants' continuous consideration for the weather when organising day-to-day activities. While these strategies were seen as important for managing comfort Darwin participants also conceded that perpetual comfort throughout the day was not feasible. But by making adjustments, less-than-comfortable conditions could be eased. By utilising a wide variety of personal adjustments, climate-appropriate property features and strategic activity scheduling, participants were able to continue with daily life without being confined to air-conditioned spaces.

The prospect that societies may be losing their vernacular resource-efficient strategies required to adjust to environmental variation without air-conditioning appears premature. Participants in both study areas employed a range of strategies that limited, even avoided, the need for air-conditioning. Nevertheless, two participants, Amelia (Melbourne) and Geoff (Darwin), expressed preferences for air-conditioning in everyday contexts, and in Geoff's case, a lack of knowledge and practice of resource-efficient strategies. Their accounts signify the possibility of broader resource-inefficient trends occurring in both study areas. To intercept everyday air-conditioning dependence, reduce energy demands during uncomfortable weather conditions, while maintaining local comfort understandings and existing adaptive capacities and outdoor engagements, more work on everyday weather relations is required. Importantly, effective resource-efficient response strategies for remaining comfortable need to be identified, supported and promoted, while infrastructural and cultural barriers of these practices need to be redressed.

Avoiding uncomfortable futures: rethinking comfort and discomfort

By turning attention from air-conditioned spaces to everyday settings, inclusive of outdoor spaces, participants showed a willingness to contend with environmental variation utilising resource-efficient adjustments. While participants identified specific weather conditions they found comfortable, they acknowledged that perpetual comfort was unrealistic. This acceptance was illustrated through weather descriptions such as 'okay', 'bearable', or 'endurable'. These accounts draw attention to restrictive language used in air-conditioning discussions. The preoccupation of thermal comfort research with (re)diversifying 'comfort' appears to have unintentionally entrenched a comfort/discomfort dichotomy. This includes social and cultural research that, while successful in their endeavours, has reproduced terminology in order to engage in engineering dialogues and encourage the loosening of standards and practices (see for example Chappells and Shove 2005).

The comfort/discomfort dichotomy is problematic, as the term 'comfort' carries connotations of the ideal and optimal. While thermal comfort is calculated on buildings occupants' level of 'satisfaction', Brager and de Dear (2003: 178) explain that:

> ... the methods for defining a 'comfort zone' or 'comfort range' of acceptable temperatures are based on associating ideal conditions only with a feeling of neutrality, or of being totally unnoticeable.

For example, ASHRAE's (2013) standard 55 and CIBSE (2006) Guide A utilise a seven-point scale to evaluate occupant comfort. On this scale, comfort encompasses 'neutral', 'slightly warm' and 'slightly cool'. Excluded from the 'comfort zone' are 'warm', 'hot', 'cool' and 'cold' sensations. Brager and de Dear (2003: 178) state 'this doesn't leave much room for anything between 'neutrality' and 'misery''. Early 'neutral' temperatures have broadened over the last few decades as standards have acknowledged the role of clothing and elevated air speeds in influencing comfort (ASHRAE 2013) and adaptive thermal comfort models have applied a wider understanding of 'satisfaction', inclusive of 'preferred' conditions (de Dear and Brager 1998). However, these discussions continue to revolve around thermal *comfort*.

If public, building and research discourses only ever discuss satisfactory indoor conditions in terms of 'comfort', 'neutrality' and 'preferred conditions', any attempt to re-diversify thermal comfort is substantially restricted. To be successful in expanding narrow indoor thermal conditions and reducing resource-dependence, research needs to move outside the 'comfort zone' altogether. Rather than discussing perfected environments through the term 'comfort', a reemphasis of 'satisfactory' conditions would assist in broadening building practices and the expectations of occupants. Furthermore, in refocusing attention from 'comfortable' to 'satisfactory' conditions, mid-ground or 'manageable' conditions that prescribe adjustments should be recognised and promoted as acceptable building environments.

To assist the process of moving away from 'comfort' discussions, Tadaki *et al.* (2014: 407) suggest building and research discourses need to understand, reconfigure and make known the cultural pursuits building practices are trying to reproduce:

> If we want to move toward a socially and environmentally sustainable architecture, perhaps it is worthwhile questioning the framing of applied climate pursuits (i.e. thickness of walls, functions of home, relations to activities) rather than treating the atmosphere as a resource to be optimised. These frames are imbued with (cultural) values and normative assumptions about the worlds we ought to make.

Similarly, O'Brien (2012: 669) and Noble *et al.* (2013) discuss the importance of 'transformative' rather than simply 'incremental' adaptations, to question our 'current systems and paradigms' that have created the climate change problems we now face.

Refocusing attention towards 'satisfactory' conditions, inclusive of both comfortable and manageable conditions, should not be perceived as progress in reverse or a decline in social standards. As Head and Gibson (2012: 709) explain:

> Modernity's relentless growth-oriented track to the future has created a situation where stepping off that track is construed as going backwards ... Some 'backward' steps are in fact entirely sensible and forward-thinking – correctives to the absurdities that creep into everyday habits, as well as a challenge to the invasion of corporate profit-making into the moments and spaces of everyday life.

If air-conditioning trends are to be intercepted, the implications associated with recent concepts of comfort need to be made clear and redirected towards more sustainable understandings and practices.

Conclusion

Experiences of physical comfort are diverse and complex. As this chapter has shown, comfort is dependent on physiological and psychological factors, including illness, fatigue, changes in oestrogen levels, normal physiological functions, residential history, activity association, as well as activities undertaken. Moreover, everyday comfort resulted not only from degree of temperature, but temperature duration, humidity, wind, rain, cloud cover and sunshine. The breadth of conditions deemed comfortable and acceptable expanded with the availability of vernacular weather response strategies, notably personal adjustments, climate-appropriate property designs and strategic activity scheduling. Due to the intricacies of comfort, differences were observed not only between individuals, but within individuals' own comfort levels. These insights contribute to accumulating evidence discrediting the necessity of 'comfort zones' provided by air-conditioning.

To continue expanding resource-intensive comfort standards, research must persist with endeavours to understand and acknowledge the diversity and complexity of comfort. This includes focusing on weather elements other than temperature, to grasp a greater comprehension of the experiences and comforts encountered outdoors and the broad range of sensations individuals come to accept and tolerate. Equally important will be sustained research into vernacular weather response strategies. Many strategies will be transferable between spaces, while others work to underscore the stringency of existing social norms that need revising. But stretching the 'comfort zone' and expanding knowledge of weather response strategies may not be sufficient to intercept the proliferation of air-conditioning. Diverting rampant trends may require abandoning pursuits for 'comfort' and advocating for 'satisfactory' and 'manageable' conditions. In doing so, the future climate change may not appear so uncomfortable.

Note

1 Temperature ranges are based on office buildings and spaces of 'moderately elevated activity' (1.1 met) where formal corporate attire (1.0 clo) is worn with humidity ratios no higher than 0.012 (ASHRAE 2013: 4).

References

ASHRAE (2013). *ASHRAE Standard-55 Thermal Environmental Conditions for Human Occupancy.*Atlanta, American Society for Heating, Refrigerating and Air-Conditioning Engineers.

Australian Bureau of Statistics (2011a). *Environmental Issues: Energy Use and Conservation* (4602.0.55.001). Canberra, Australian Bureau of Statistics.

Australian Bureau of Statistics (2011b). *Basic Community Profile: 2011 Census of Population and Housing: Greater Melbourne* (2001.0). Canberra, Australian Bureau of Statistics.

Brager, G. and de Dear, R. (1998). 'Thermal adaption in the built environment: A literature review.' *Energy and Buildings* 27(1): 83–96.

Brager, G. and de Dear, R. (2003). 'Historical and cultural influences on comfort expectations.' *Buildings, Culture and Environment: Informing Local and Global Practices.* R. Cole (Ed.) Oxford, Blackwell: 177–201.

Bureau of Meteorology (2008). *Climate of Australia.* Canberra, Australian Bureau of Meteorology.

Bureau of Meteorology (2014a). *Monthly climate statistics for 'Darwin Airport'* [014015]. Canberra, Australian Bureau of Meteorology.

Bureau of Meteorology (2014b). *Monthly climate statistics for 'Melbourne Regional Office'* [086071]. Canberra, Australian Bureau of Meteorology.

Cândido, C., de Dear, R., Lamberts, R. and Bittencourt, L. (2010). 'Cooling exposure in hot humid climates: Are occupants "addicted"?' *Architectural Science Review* 53(1): 59–64.

Chan, C. and Ryan, D. (2009). 'Assessing the effects of weather conditions on physical activity participation using objective measures.' *International Journal of Environmental Research and Public Health* 6(10): 2639–2654.

Chappells, H. and Shove, E. (2005). 'Debating the future of comfort: Environmental sustainability, energy consumption and the indoor environment.' *Building Research and Information* 33(1): 32–40.

CIBSE (2006) *Environmental Design: CIBSE Guide A*. London: CIBSE.

City of Melbourne (2011). 'Melbourne's Weather and Climate.' http://www.melbourne.vic.gov.au/AboutMelbourne/MelbourneProfile/Pages/Melbourneweatherandclimate.aspx, viewed 29 September 2014.

Cooper, G. (2008). 'Escaping the house: Comfort and the California garden.' *Building Research and Information* 36(4): 373–380.

de Dear, R. and Brager, G. (1998). 'Thermal adaptation in the built environment: A literature review.' *Energy and Buildings* 27(1): 83–96.

de Vet, E. (2013). 'Exploring weather related experiences and practices: Examining methodological approaches.' *Area* 45(2): 198–206.

de Vet, E. (2014). *Weather-Ways: Experiencing and Responding to Everyday Weather*. PhD thesis, University of Wollongong, Wollongong.

de Vet, E. (2017). 'Experiencing and responding to everyday weather in Darwin, Australia: Implications of inherent practices of climate change adaptation.' *Weather, Climate and Society* 9(2): 141–154.

Fuller, S. and Bulkeley, H. (2013). 'Changing countries, changing climates: Achieving thermal comfort through adaptation in everyday activities.' *Area* 45(1): 63–69.

Gram-Hanssen, K. (2008). 'Consuming technologies – Developing routines.' *Journal of Cleaner Production* 16(11): 1181–1189.

Hajat, S., O'Connor, M. and Kosatsky, T. (2010). 'Health effects of hot weather: From awareness of risk factors to effective health protection.' *Lancet* 375(9717): 856–863.

Haldi, F. and Robinson, D. (2010). 'On the unification of thermal perception and adaptive actions.' *Building and Environment* 45(11): 2440–2457.

Head, L. and Gibson, C. (2012). 'Becoming differently modern: Geographic contributions to a generative climate politics.' *Progress in Human Geography* 36(6): 699–714.

Healy, S (2008). 'Air-conditioning and the "homogenization" of people and built environments.' *Building Research and Information* 36(4): 312–322.

Hitchings, R. (2009). 'Studying thermal comfort in context.' *Building Research and Information* 37(1): 89–94.

Hitchings, R. (2010). 'Urban greenspace from the inside out: An argument for the approach and a study with city workers.' *Geoforum* 41(6): 855–864.

Hitchings, R. (2011). 'Coping with the immediate experience of climate: Regional variations and indoor trajectories.' *Wiley Interdisciplinary Reviews: Climate Change* 2(2): 170–184.

Hwang, R.-L., Cheng, M.-J., Lin, T.-P. and Ho, M.-C. (2009). 'Thermal perceptions, general adaptation methods and occupant's idea about the trade-off between thermal comfort and energy saving in hot-humid regions.' *Building and Environment* 44(6): 1128–1134.

Indraganti, M. (2010). 'Behavioural adaptation and the use of environmental controls in summer for thermal comfort in apartments in India.' *Energy and Buildings* 42(7): 1019–1025.

Indraganti, M. and Rao, K. D. (2010). 'Effect of age, gender, economic group and tenure on thermal comfort: A field study in residential buildings in hot and dry climate with seasonal variations.' *Energy and Buildings* 42(3): 273–281.

Kosonen, R. and Tan, F. (2004). 'The effect of perceived indoor air quality on productivity loss.' *Energy and Buildings* 36(10): 981–986.

Kwok, A. G. and Rajkovich, N. B. (2010). 'Addressing climate change in comfort standards.' *Building and Environment* 45(1): 18–22.

Levermore, G. J. (2008). 'A review of the IPCC assessment report four, part 1: The IPCC process and greenhouse gas emission trends from buildings worldwide.' *Building Services Engineering Research and Technology* 29(4): 349–361.

Lin, T.-P. (2009). 'Thermal perception, adaptation and attendance in a public square in hot and humid regions.' *Building and Environment* 44(10): 2017–2026.

Maller, C. J. and Strengers, Y. (2011). 'Housing, heat stress and health in a changing climate: Promoting the adaptive capacity of vulnerable households, a suggested way forward.' *Health Promotion International* 26(4): 492–498.

Meyer, W. (2000). *Americans and Their Weather*. New York, Oxford University Press.

Nicol, F., Humphreys, M. A. and Roaf, S. (2012). *Adaptive Thermal Comfort: Principles and Practice*. Oxon, Routledge.

Nicol, J. F., Raja, I. A., Allaudin, A. and Jamy, G. N. (1999). 'Climatic variations in comfortable temperatures: The Pakistan projects.' *Energy & Buildings* 30(3): 261–279.

Nikolopoulou, M. and Steemers, K. (2003). 'Thermal comfort and psychological adaptation as a guide for designing urban spaces.' *Energy and Buildings* 35(1): 95–101.

Noble, I., Huq, S., Anokhin, Y., Carmin, J. A., Goudou, D., Lansigan, F., Osman-Elasha, B. and Villamizar, A. (2013). 'Adaptation needs and options.' *Climate Change 2014: Impacts, Adaptations and Vulnerabilities*, Contributions of the Working Group II to the Fifth Assessment Report of the Intergovernmental Panel on Climate Change, Cambridge, Cambridge University Press: 1–51.

O'Brien, K. (2012). 'Global environmental change II: From adaptation to deliberate transformation.' *Progress in Human Geography* 36(5): 667–676.

Parish, K (2007). 'Paradise crushed under white shoe heels?' Club Troppo http://clubtroppo.com.au/2007/02/09/paradise-crushed-under-white-shoe-heels/, viewed 29 September 2014.

Pearlmutter, D. (2007). 'Architecture and Climate: The Environmental Continuum.' *Geography Compass* 1(4): 752–778.

Roaf, S., Nicol, F., Humphreys, M., Tuohy, P. and Boerstra, A. (2010). 'Twentieth century standards for thermal comfort: Promoting high energy buildings.' *Architectural Science Review* 53(1): 65–77.

Rothwell, N. (2007). 'Down disconsolate streets.' *The Australian*, 7 February: p14.

Sahakian, M. D. and Steinberger, J. K. (2011). 'Energy reduction through a deeper understanding of household consumption: Staying cool in Metro Manila.' *Journal of Industrial Ecology* 15(1): 31–48.

Shove, E. (2003). *Comfort, Cleanliness and Convenience: The Social Organisation of Normality*. Oxford: Berg.

Stinson, S., Bogin, B. and O'Rourke, D. (2012). *Human Biology: An Evolutionary and Biocultural Perspective*. New Jersey, USA: Wiley-Blackwell.

Strengers, Y. (2008). 'Comfort expectations: The impact of demand management strategies in Australia.' *Building Research and Information* 36(4): 381–391.

Strengers, Y. (2010). 'Air-conditioning Australian households: The impact of dynamic peak pricing.' *Energy Policy* 38(11): 7312–7322.

Strengers, Y. (2012). 'Peak electricity demand and social practice theories: Reframing the role of change agents in the energy sector.' *Energy Policy* 44: 226–234.

Strengers, Y. and Maller, C. (2011). 'Integrating health, housing and energy policies: Social practices of cooling.' *Building Research and Information* 39(2): 154–168.

Tadaki, M., Salmond, J. and Le Heron, R. (2014). 'Applied climatology: Doing the relational work of climate.' *Progress in Physical Geography* 38(4): 392–413.

Tucker, P. and Gilliland, J. (2007). 'The effect of seasons and weather on physical activity: A systematic review.' *Public Health* 121(12): 909–922.

Williamson, T. J., Coldicutt, S. and Penny, R. E. C. (1989). *Thermal Comfort and Preferences in Housing: South and Central Australia*. Adelaide, S.A, Department of Architecture, The University of Adelaide.

4 (Re)creating a sense of comfort

Post-disaster homemaking

Stephanie Harel

When a major flood is described as a 'once in 100 years' event, then by definition it is something extraordinary. In reality, it was much less than 100 years after 1974 that the next big flood came to Queensland's capital city. On January 13th, 2011, the Brisbane River peaked, at 4.46 metres, flooding approximately 22,000 across the city (van den Honert and McAneny 2011). For those affected, this date marked the moment where their lives were forever altered.

In January 2011, Liz[1] watched in horror as floodwater consumed her post-war cottage in Brisbane's West End:

> From the safety of my friends' couch, it all seemed a bit surreal, like it was happening somewhere else, to someone else and not me.

A few days after the floodwater had receded, Liz was able to return to her house to survey the flood damage. Devastatingly, the floodwater had reached just below her ceiling, destroying almost everything she owned:

> When the reality hit, and I realized that my life, everything I'd worked so hard for…that so much has been destroyed…it was just horrible…It was heartbreaking.

The days that followed '*ran together like a rush of adrenaline*'. In this immediate aftermath, Liz spent her days cleaning and removing the thick layer of mud that coated her family home. With the help of '*kind strangers*', whose support offered her '*comfort*' at a difficult time, Liz gradually began to re-establish her domestic space. Slowly, with the presence of familiar objects, her house began to feel like home again:

> To come back and just see a shell, wasn't good. Because it was all different. It just wasn't mine – you know. But then after Graham put the curtains up and he put some photos on the wall, it started to feel – you know. Now I feel more comfortable in my bedroom and living room. I hate the study because there's nothing. It's just empty. The bed's gone

> and photos and so forth. But after the things came back and had been rescued and cleaned – so that I feel content sitting in the living room. I guess that's why I didn't want to get rid of the sofa suite, because it was something I knew. That was my feeling coming back. It just didn't feel right. And the floor was different. When I looked at this I was like 'ahhh!' it was so garish and yellow. But I've got used to it. When I started getting a few things around me again, you know – that's when it just felt right again.

Through the impact of the crisis, Liz's flood-damaged home was transformed into a space of discomfort and anxiety. Her domestic space became somewhere she barely recognized and struggled to form an attachment to. In this sense, the floods played a large part of her home *unmaking* (Brickell and Baxter 2014). Liz's narrative also, however, begins to suggest the importance of familiar objects in the post-disaster environment. By bringing these objects back into her damaged house, Liz was able to recreate a sense of home in the aftermath of crisis.

This chapter builds on Liz's story, analyzing the narratives of displaced disaster victims to demonstrate the capacity of natural forces to reconfigure human and domestic geographies of comfort. Bissell (2008) presents a multidimensional understanding of comfort in terms of an objective capacity, an aesthetic sensibility and an affective resonance. In this chapter, I understand a sense of comfort that strengthens, supports and consoles in the aftermath of tragedy. More specifically, the purpose of this chapter is to explore how disaster victims negotiate and (re)create a sense of comfort in an often uncomfortable and exclusionary post-disaster landscape. Considering ideas of domestic materiality (Miller 2008), I will demonstrate the capability of familiar objects to offer comfort in the aftermath of disaster.

The concept of place is central to this enquiry, and while I will not reflect specifically on place literature in this chapter, I do acknowledge an emotio-spatial hermeneutic. As Davidson and Milligan (2004, 524) suggest, 'emotions are understandable – "sensible" – only in the context of particular spaces. Likewise, place must be *felt* to make sense'. In other words, meaningful senses of space emerge only via movements between people and places. This chapter is informed by recent work in emotional geography (Pile 2010, 2011) that focuses on the emotions experienced by embodied individuals and attaching to particular places, pointing to the interconnectedness of people and their environments. Ultimately, this work usefully shows how emotions are integral to how places are imagined and portrayed. This has profound implications for understanding the embodied experiences of the post-disaster environment.

Disasters and home

Home is both material and affective. Not merely a physical locality, home is a milieu of shifting cultural associations and 'ideal' meanings

(Gorman-Murray 2007; Blunt and Dowling 2006; Easthope 2004; Young 1997). While home is commonly associated with the site of the house in contemporary Western society, the affective capacity of home cannot be reduced to the physical space of a house. Indeed, as Gorman-Murray (2007, 229) suggests, 'For a house to become a home, it must be imbued with a range of meanings, feelings and emotions'.

The early house as haven theories idealize home as a symbolic site of rootedness, belonging, comfort and safety (Moore 2000; Guiliani and Feldman 1993). More recently, however, geographers have begun to make a more notable commitment to the exposing large-scale disruptions to home (Brickell and Baxter 2014; Brickell 2012; Schroder 2006), for example through the wrath of natural disasters. These studies of 'extreme geography' (Brickell 2012) explore how a sense of home and disruptions to that attachment are discussed in a number of recent disaster studies (Gorman-Murray 2014 et al. 2014; Morrice 2013; Hawkins and Mauer 2011). These studies suggest that when locations are disrupted by disaster, so too are the individual and social identities that are tied to these specific places. Loss and/or damage to home and neighbourhood sever the routine and normalcy of one's social and material environment, disrupting ontological security.

While a widespread definition remains disputed, I understand natural disasters as social phenomenon (Brun and Lund 2008). These catastrophic events occur within a society, causing both physical and social disruption. This might include damage to housing and infrastructure as well as communities and social relations. Disasters are also events that are grounded in the very idea of home. Commonly associated with loss of home and the challenges associated with (re)creating a sense of home in often unfamiliar and exclusionary landscapes, they highlight both the porosity of home and the formal and cultural politics of home.

Catastrophic events also remind us of the power of nature and the fragility of contemporary society. As a researcher, this reality was profoundly reinforced as I spent time in New Orleans, Louisiana in the aftermath of 2005's Hurricane Katrina. As I navigated the still-ravaged neighbourhoods of the Lower Wards, destroyed houses lined the streets. My work in post-disaster communities draws further attention to the reality of a complicated post-disaster recovery process and disruption to the many scales of home.

It is evident, then, that natural disasters have significant impact upon domestic geographies of the familiar. They have the ability to disrupt and alter how we perceive and relate to domestic spaces (Morrice 2013; Blunt and Dowling 2006). Kaika (2004) develops this argument, considering how familiar domestic objects can also behave in unusual ways in the aftermath of crisis (Kaika 2004). Such incidents produce a feeling of uneasiness, discomfort and anxiety, which threatens to tear down the security and familiarity of domestic spaces (Vidler 2002). This exposure to the limits of domestic bliss ultimately can lead to a feeling of, 'not being at home in one's home' (Vidler 1992, 4).

Current research in geography also focuses on the concept of home *unmaking* (Brickell and Baxter 2014; Fernandez Arrigoitia 2014). As Brickell and Baxter (2014, 134) explain, 'Home unmaking is the precarious process by which material and/or imaginary components of home are unintentionally or deliberately, temporarily, or permanently, divested, damaged or even destroyed'. By this definition, then, natural disasters are particularly visible examples of home unmaking.

Fernandez Arrigoitia (2014) usefully illustrates the impact of home unmaking practices in Puerto Rican apartment blocks. She explores how materials on the edge of home, such as a decaying staircase or failing lifts prevent elderly residents from venturing outside, and act as a reminder of an unpleasant past. She notes that in circumstances, however, residents still attempt to 'make home' in this space. Despite living in difficult, uncomfortable spaces, the routines and rhythms of life go on. These processes, which unfold into the present, can also reach into the past (Fernandez Arrigoitia 2014). This temporal connection is particularly relevant of the post-disaster home. Part of the reason why the destruction of home is so traumatic is that the present is thick within the memories of happier times and relationships. Trauma is not necessarily contained in an event, but in the way this event is experienced. Indeed Traumascapes are full of visual and sensory triggers and capable of eliciting a whole palette of emotions. In this sense, they can catalyze and shape remembering and reliving of traumatic events (Tumarkin 2005).

The materiality of the post-disaster home

Above I have suggested that part of the reason why loss of home as a result of disaster is so traumatizing is that domestic spaces are extremely important in our daily lives. This is because they reflect and shape our identities (Blunt and Dowling 2006; Miller 2001). This focus on domestic spaces, however, does not detract from the understanding that one can be 'at home' on many scales (Brickell 2012; Blunt and Dowling 2006; Blunt and Varley 2004; Tuan 1977). As Blunt and Dowling (2006, 23) point out:

> Home does not simply exist, but is made. Home is a process of creating and understanding forms of dwelling and belonging. This process has both material and imaginative elements. Thus people create home through social and emotional relationships. Home is also materially created – new structures formed, objects used and placed.

Csikszentmihalyi and Rochberg-Halton (1981) argue that domestic objects comprise the endogenous being of the owner. This is because these objects are chosen by the owner and therefore reflective of their own identity. Other scholars enhance this argument (Marcoux 2001; Pearce 1992; Appadurai 1986) acknowledging that objects are entangled into many aspects of our social and individual lives, and indeed play a considerable role in identity construction.

The importance of material culture within domestic space is highlighted by research that demonstrates its centrality in homemaking and creating a sense of home (Putnam 1993). Indeed, over the past decade, domestic geographies have undergone a 'material turn', as academic work acknowledges both the emotional and material aspects of homemaking (Liu 2014). Material geographies recognize the importance of objects within the home as critical in the formation of new identities, shaping memories and scales of identification (Brickell and Datta 2011; Datta 2008). Objects are, as Whitmore (2001, 59) succinctly points out, 'embedded with the essence of life's experiences'.

Contemporary studies of both homed and homeless people (see, for example, Miller 2010, 2008) continue to explore how objects play a key role in the (re)creation of home. Objects, for example, act as inducers of reminiscence (Casey 2009) and possess the ability to arouse a reminiscent frame of mind. Extant research confirms the importance of cherished possessions (Ureta 2007; Whitmore 2001), which are more than a mere memento. Instead, these items offer an evocation of one's life and experiences. As Sherman and Dacher (2005, 72) contend, '[it is the] objects themselves that lend meaning to the dwelling. Through the objects it contains, the space is imbued with significance. This is a dynamic and non-linear process'. Cherished objects are therefore experienced as meaningful, rather than merely *known* to be meaningful. Ahmed (2000, 89) supports this argument, suggesting, 'The lived experience of being-at-home...involves the enveloping of subjects in a space which is not simply outside them: being-at-home suggests that the subject and space leak into each other, *inhabit each other'*. Home is, in this sense, the lived experience of locality.

For bodies on the move, material culture is perhaps even more significant in terms of homemaking. This is because objects are more easily transported than buildings or neighbourhoods. By transporting selected objects from place to place, migrants are, as Shumaker and Conti suggest, 'able to recreate home in each new setting' (1985, 249). In terms of the context of this chapter, disaster displacement separates victims from their material culture and migrants are often left to negotiate a home for themselves in a situation where they are forced to leave home repeatedly. In this sense, it is worthy to consider whether the disaster-displaced might withdraw their emotional attachments from the physical structure of 'home', and instead focus their energies on mobile objects that can be carried along and are related to 'home' in a conventional sense (Montasim and Heynen 2011; Hooper-Greenhill 2000). As Wise (2000, 229) argues, therefore:

> Home can be a collection of objects, furniture and so on that one carries with one from move to move. Home is the feeling that comes when the final objects are unpacked and arranged and the spaces seem complete.

For the disaster-displaced, objects offer a reminder of who they are, where they have come from and the journey they have endured. These objects can include pieces of personal symbolism to them, but also include objects that might seem mundane and trivial (Walsh 2006). Photos, for example, can visually represent a domestic memory, and a powerful emotional connection with a place (Walsh 2006; Rose 2003). Ahmed (2004) explores this further through a relational understanding of emotion. Drawing on this body of work, this chapter further considers not only love between and among individuals in the aftermath of crisis, but also the love between individuals and objects.

Bodies are, as Ahmed (2008) suggests, shaped by contact with objects and with others. To be affected by an object in a good way is to have an orientation towards an object as being good (Ahmed 2008). Being 'orientated' is therefore synonymous with feeling happy or 'at home' and having certain objects within reach (Ahmed2006). In this sense, we move towards and away from objects depending on how we are affected by them (Ahmed 2008). Ahmed considers these arguments further, through the experience of alienation. She suggests that when we feel pleasure from objects, we feel aligned. In the same way, we feel alienated when we do not experience pleasure from proximity to objects that are regarded as being good. In this sense, familiar objects can offer a source of comfort to bodies that move. In an unfamiliar environment, these objects affect us in a positive way (Ahmed 2008).

Brisbane and the 2011 floods

I opened this chapter with a small excerpt taken from an interview I conducted in Brisbane in 2012. This interview formed part of my doctoral project, which used the 2011 Brisbane floods and the 2011 Christchurch earthquake to explore the emotional dynamics of the post-disaster environment. More explicitly, this project explored people's emotional responses to disaster, highlighting the impact of crisis on domestic geographies of the familiar.

For the purpose of this chapter, I will draw specifically from the semi-structured interviews I conducted with residents who were displaced from the Brisbane floods. Part of my interview processes involved asking participants to share an 'important object' from their displacement experience. I left this request deliberately vague, in order to encourage a wide range of objects. This method allowed an opportunity to understand how perhaps taken-for-granted everyday objects play a part in the (re)creation of home for displaced disaster victims. In the analysis that follows, I will focus on my interview narratives to demonstrate how the floods can transform a once familiar and comfortable domestic space, into a space of discomfort and unfamiliarity. In light of this, I also reflect on how material objects contribute to the (re)establishment of a comfortable home in the aftermath of crisis.

The post-disaster 'home': an uncomfortable space?

The narratives of displaced flood victims explicitly demonstrate the underlying trauma associated with life after disaster. The Brisbane floods caused a sudden and devastating disruption to the lives of city residents. This emotional impact was felt both in the immediate aftermath of the event and during the months and years that constitute the post-flood recovery period. My 2011 interview with Kim clearly demonstrates the reality of this trauma. Despite severe flood warnings, Kim did not evacuate from her home in East Ipswich. The day after the river peaked, she, therefore, had to be rescued by a boat. Her home, once a place of comfort and familiarity, was left barely visible – submerged under 10 feet of water. It did not take long before the reality of her loss began to sink in:

> And then we woke up and it was a world of hurt, you know? Like it all hits you that everything's gone…it's been really hard ever since.

During my fieldwork, displaced residents all expressed an emotional longing for their domestic space. Despite her deeply traumatic ordeal, Kim still, for example, expressed a desire to return to her flood-damaged bungalow: *'I was hopeful…excited even to get back!'* For Kim, her residence was understood as a locus of positive experience. In this sense, home is understood as a place of memory; a safe and comfortable haven from the threatening problems of the post-disaster environment.

As John, a 56-year-old resident of Goodna explained his own view of home: 'Your home is usually – what is it – it's the key place that you go to as your refuge so to speak. Your sanctuary'. The nostalgic and romanticized view John holds towards his domestic space drives his desire to return home after displacement. These feelings of happiness, comfort and belonging construct a static vision of home, as somewhere that must be returned to. Anelle, a lifelong Brisbane resident, returned to Goodna a few weeks after her initial displacement. She told me: 'even if we have to rebuild again and again, we will. There are too many memories, this is home and there's nowhere else I would rather be'.

My narratives, however, also provide a deeper perspective of the negative aspects of people–place relationships. The unforgiving post-flood landscape can become a frightening place – a place where memories and anticipation collide with the reality of returning to a home drastically altered by disaster. Many residents, for example, explained how they were initially confronted with an unfamiliar domestic space in the aftermath of the floods: 'It didn't even look like our house, just a strange shell I didn't know'. The painful emotions that resonate through my narratives powerfully highlight the tensions and ambiguities that victims of disaster face as they negotiate the reality of rejecting the home they romanticized at a distance. Returning to a place in reality can be deeply disorientating when this romanticized image of home is not met. 36-year-old Janine, for example, discussed her disappointment upon returning to view her one-story bungalow in Brisbane's West End. She told me:

> I felt devastated. I thought everything would be okay but I was so disappointed at what I actually saw. When I walked in, there was this very thick mud. It was a horrible experience to see things that meant a lot to you, damaged and broken.

Sally, a 40-year-old mother of two shared a similar story. At the time of the floods she was living in Indropilly with her two young boys. The negative emotions she experienced when present at her flood-damaged property were a key influence on her decision to move away from the suburb:

> I feel sick in the stomach every time I go there. I do. It's just instant. I sort of – there's a few times I thought, oh what's going on, you know? I feel really queasy. And then I worked out, it's just every time I go there. Just being there and seeing it. Seeing our house, you know, like that. And the dirty muddy water, the idea of that. That just does it. You think, you're better off somewhere else.

For those who, like Sally and Janine, experience the residence as a source of trauma, they feel alienation towards the place that was once one of their most intimate and sacred spaces. In this sense, their permanent return to an uncomfortable place they no longer feel an attachment to, is too emotionally distressing.

It is clear, then, that flood victims face an extremely difficult time in the aftermath of disaster. For those who face disappointment upon returning to their flood-damaged home, this trauma is exacerbated. This understanding raises important questions in terms of post-disaster homemaking – what facilitates this process for those who find themselves struggling to deal with the reality of an altered domestic space? The following sections will explore this question in more detail.

Comforting objects

The loss of not just a house, but also a home is one of the most devastating impacts of a natural disaster (Wilford 2008). The central consequence of the materiality of the flood-transformed house is one of alienation and appropriation. The loss in this context is unique in that much of what has been 'lost' is still present. Many of the residents returned to their flooded homes to find everything still physically present, but covered in a thick layer of mud or damaged by water. Salvaging objects from the interior of their homes is a way for flood victims to reclaim this space. Indeed, reclaiming these objects can also be understood as reclamation of the self (Taylor 2006).

During my fieldwork in Brisbane, participants brought a number of interesting objects to interviews. These seemingly trivial objects might not hold any meaning to others, but in this environment they become clearly emotionally significant to flood-affected individuals (Tolia -Kelly 2004). This is because they have the ability to transform shelters, temporary accommodation, or indeed new houses, into something that resembles home (Wilford 2008). The movement of objects also creates a physical and emotional continuity for flood victims as they re-establish their lives following disaster (Hooper-Greenhill 2000).

Each interview I conducted explored these emotions in an intimate way. In January 2012, I interviewed Laura. Before the floods, she was living in Chelmer with her husband and young daughter. During our interview, Laura discussed the emotional challenges she faced in the aftermath of disaster. Her interview also highlighted the significance of material objects in this post-disaster environment:

> One of the lessons in the aftermath is that every step back to normality – such as having a house, a roof, or a job, is a cause for celebration – even if this step back means several steps away from the area. It is an empowering feeling to move out of a temporary 'in-limbo' situation into a confirmed future direction. And it is comforting to take a little of that old life with you, whether in the form of a photo or a day bed.

After the flood, Laura and her family relocated to East Ipswich. As her interview suggests, keeping familiar objects with her allowed her to recreate a sense of 'home' in a new physical dwelling. The presence of these objects offered comfort and familiarity during a time of uncertainty and anxiety.

Narratives also demonstrate how the presence of familiar objects and the act of organizing one's home in the same way it looked before the floods help to recreate a sense of familiarity and comfort in the post-disaster domestic space. For example, Megan, a 56-year-old who was living in Graceville at the time of the floods, discussed how she began to feel more comfortable once her flood-damaged house was full of her belongings again: 'Unpacking boxes of my belongings was a milestone in my recovery. Having my own things around me made it feel like home again'. Through the presence of familiar material objects, Megan's Graceville bungalow slowly began to feel like home again. Here homemaking is an active process, comprising both imaginative and material elements.

Throughout my fieldwork, photographs were the most common objects participants shared with me during interviews. These photographs facilitate and represent the transport of home across both time and space. They form a virtual thread temporally and geographically between locations (Rose 2012, 2003). In a disaster situation, photographs are often the last thing a person might grab before they leave home and the first thing they go back to look for. For many, returning to find photographs destroyed by floodwater was deeply emotionally distressing. Rose, for example, told me: 'To go back and see all my photos and albums destroyed in such a way was heartbreaking. They're the things you can never get back'. Many research participants purposely took photographs with them when they left their houses before the floodwater peaked. Betty explained: 'Since I'd been through the 1974 flood, I knew things like photos were not salvageable. So I packed up many as I could, and took them with me'.

Flood-affected individuals also explicitly highlighted the value and comfort of having photographs with them in the aftermath of the floods. Leanne produced a photograph of her and her husband (Figure 4.1).

During our interview, she treated the snapshot with care and tenderness:

> I brought this photo because it's something special to me. Something I can look at again and something I can remember from before. Something that wasn't scarred from that day in January, when absolutely everything in my life changed, or was in some way destroyed.

While Leanne and her family stayed in temporary accommodation following the floods, this photograph became, 'a symbol of hope. Something that made us feel more at home while we waited to return to Ipswich'. Leanne's narrative demonstrates the way in which displaced flood victims place emotional attachments onto material objects in the aftermath of disaster. For her, this family photograph represents a specific domestic memory. Keeping this photo close while living in temporary accommodation helped create a sense of home in an unfamiliar setting. As Ahmed (2008) suggests, bodies are shaped by contact with objects and others. Leanne felt 'at home', or 'orientated' because she had this photograph within reach. Her narrative, therefore, shows how objects offer comfort to bodies that move.

Figure 4.1 '[This photo] is a symbol of hope' (Leanne).

Other participants shared similar stories. Catherine, for example, verbally acknowledged how mundane objects, such as photographs, can hold heightened meaning in the post-disaster environment. During our interview, she presented a family photograph she managed to salvage from the flood damage:

> This is one of the few photographs we have from before the floods. A lot of our other photos were destroyed in the mud and floodwater, but we managed to salvage this one. I've always liked it... It's funny how the little things become especially important during a time of crisis.

In the post-disaster environment, Catherine places a strong emotional attachment onto this particular photograph. Her narrative also describes how this photo gives her comfort in her new house in Sherwood:

> It's weird trying to start again somewhere new and even though we feel settled now, it's very hard to pick yourself up and move on after you've lost everything. Having little things that were familiar to me made that transition just that little bit easier. They made me feel more comfortable.

Again, this interview demonstrates how material possessions, which might appear mundane in everyday life, play a vital and influential role in post-disaster homemaking practices. They hold emotional attachments and provide comfort to disaster victims during a harrowing time.

In addition to photographs, other objects participants shared with me included artwork, jewelry and kitchenware. In 2011, I interviewed John at the Mount Ommaney community centre. He attended this interview with a simple canvas painting. While he verbally acknowledged that this painting, 'might not mean anything to anyone else', it offers him 'comfort' during the aftermath of the flood. He went on to explain this further, suggesting his attachment to this material possession:

> I've always loved this piece. I think its simplicity is refreshing. When we returned to East Ipswich, one of the first things I did was hang it on the wall in our hallway. Hanging pictures around the house was quite a moment for me. Probably the moment when I really let myself believe that we were home.

Re-hanging artwork in his renovated home signified finality to John's displacement and therefore an important milestone in his recovery process. His narrative clearly demonstrates how the presence of these material objects lends meaning to his dwelling. His artwork, therefore, played an important role in his post-flood recovery, helping to recreate a sense of 'home' in his domestic space.

Two of the participants I interviewed in Brisbane shared pieces of jewelry with me. As Alison explained: 'It's been such a horrible time, that I've found comfort in the smallest things. Wearing my favourite necklaces for example'. For flood victims, the emotional post-disaster landscape is often an unfamiliar place, where much of what defines a sense of 'self' has been destroyed in a traumatic manner. During this time, victims take comfort in things that allow them to reconstitute the 'self' and their own identity. For Alison, the act of wearing jewelry is a key part of this process.

Pauline, a young, single mother living in Graceville at the time of the flood also brought a cherished piece of jewelry to our interview. After delicately placing it on the table in front of us, her narrative discussed the emotional connection she feels towards this object:

> Anything that you manage to keep or save becomes especially special. This necklace was Mum's and it was already special to me. When I found it in the debris, I cried. It's comforting to have these special things around you when you've been through something like this. These 'things' make it a little easier to get back on track.

This cherished personal possession holds a deep emotional significance to Pauline. The memories attached to this necklace provide her with a sense of continuity and comfort during the aftermath of the floods.

As depicted by my female participants, domestic routines (such as cooking and cleaning) are commonly disrupted through disaster and during the subsequent recovery period following the event. My narratives suggest that during this time, cooking becomes an important way of staying connected with home through a sensory and emotional geography. While some feminist scholars have argued that the home (particularly the kitchen) is a major site of women's oppression (see Ahrentzen 1997), this was not the case with my participants in this instance. Instead, my female participants spoke fondly of cooking, discussing the positive emotions associated with this domestic practice.

One of my participants, for example, brought a beautiful china jug to our interview (Figure 4.2). Nancy's flood experience was one characterized by a deep feeling of disconnect from her domestic space: 'I guess the overwhelming feeling I had at that moment was disconnection. From everything that was normal and real. I was heartbroken.' When reflecting on her chosen object, she told me:

> I've always felt at home in the kitchen. I love to cook for family and friends. So my kitchenware, things like this [points to jug], are very important to me. So I enjoy cooking, but ever since the floods it has become even more important to me. It's like an escape of sorts. A time when I can just forget about all the bad things that have happened to us.

Nancy's narrative suggests how food can help to create a new sense of 'home' in new settings, as it provides a connection to one's life experiences (Watson

Figure 4.2 'Kitchenware, things like this, are very important to me' (Nancy).

and Caldwell 2005; Ashley et al. 2004; Warde 1997). The emotional and affective relationship between Nancy and her kitchen also demonstrates her visceral experience with food (Longhurst et al. 2009), which helps people feel 'at home'. In the aftermath of the floods, Nancy's kitchenware has taken on a new meaning for her and her family:

> These are things that are familiar to us, and that's important during times like this. And I think because they are connected to cooking, and eating, a time for family and friends to come together, these objects have sort of become a symbol of togetherness. In the immediate aftermath, we lived temporarily with friends and I kept saying to my husband that I couldn't wait to have our own kitchen again so I could cook. Now we're in our new home, I enjoy cooking even more. And I'm happy to have some of the objects from our old life here.

Through the above narrative, Nancy demonstrates how feelings of belonging are experienced as both sited and mobile, illuminating the complex, personal and affirmative workings of bodies and emotions. Law contends that 'food acquires its meaning through the place it is assembled and eaten' (2001, 275). Everyday routines and practices, such as cooking and cleaning, also transform private space, making it distinct from any other. Essentially, they lend meaning to a dwelling, giving the house its identity as home (Tolia-Kelly 2004; Petridou 2001). Through the practice of making food and using familiar kitchen objects, Nancy is able to ascribe meaning to this new domestic space. The seemingly mundane experience of cooking becomes an expression of her own subjectivities. Her new kitchen offers a space of comfort and familiarity in a new domestic place.

Conclusions

Our contemporary society is one of heightened risk (Hyndman 2007). This expansion of fear comes from a variety of sources, including global climate change, geopolitical tensions and the threat of natural disasters. Many have documented a rising trend both in terms of the number of disasters occurring, and the economic cost of their destruction (Guha-Sapir et al. 2013). This chapter demonstrates the effect of these disasters on domestic geographies of the familiar. Through an analysis of the *experience* of disaster, I have demonstrated the capacity of natural forces to reconfigure both human and domestic geographies of comfort. Natural disasters are hugely disruptive to the emotions people feel to places, and indeed to personal levels of comfort and safety. For example, in the post-disaster context, home is often rendered uncanny. No longer a safe haven from the threatening world, it becomes a place of discomfort and anxiety for those affected by disaster. This indeed reminds us of the destructive power of nature and the fragility of contemporary society.

I would like to close this chapter by reflecting on two points. Firstly, the traumatic effects of natural disasters are complex and long-term. My research acknowledges the ongoing struggles disaster victims continue to face in the aftermath of tragedy. In this sense, much further research on the topic is needed in order to fully understand the complexities of the post-disaster environment. While this chapter begins to explore the emotional and dynamic process of post-disaster homemaking, future research would be wise to explore this process in other post-disaster contexts (such as the aftermath of fires, tornadoes and cyclones). The individuality of emotions suggests that this research will undoubtedly be extensive. Still, only with continued investigation into the emotional dynamics of natural disasters and the geographies of the post-disaster 'home' can we hope to share scientific knowledge between academic audiences, victims of catastrophic events and disaster planners, in an effort to lessen the effects of future events.

This brings me to my second point: The issues I have raised here have many practical implications for planners and policymakers who are actively involved in post-disaster recovery strategies. The importance of people's emotional responses to disaster and an understanding of how victims relate to their domestic space in the aftermath of crisis should not be neglected or downplayed. Instead, the social and emotional geographies of the post-disaster environment must be acknowledged and understood. Effective and coordinated communication between local and national emergency management organizations, NGO's and affected community members is a vital aspect of post-disaster recovery strategies. As fresh questions are raised about the social and cultural geographical dimensions of climate change, geographers would be wise to continue to our focus on disasters, human emotions and the complex connection between bodies and places.

Note

1 All of the names used in this chapter are pseudonyms.

References

Ahmed, S. (2000) Embodying strangers. In Horner, A. and Keane, A. (eds.) *Body matters: Feminism, textuality, corporeality*. Manchester: Manchester University Press, 85–97.
Ahmed, S. (2008) Sociable happiness. *Emotion, Space and Society* 1 (1), 10–13.
Ahmed, S. (2004) *The cultural politics of emotion*. Edinburgh: Edinburgh University Press.
Ahrentzen, S. (1997) The meaning of home workplaces for women. In Jones, J. P. III, Nast, H. and Roberts, S. (eds.) *Thresholds in feminist geography: difference, methodology, representation*. Lanham, MD: Rowman and Littlefield, 77–92.
Appadurai, A. (1986) *The social life of things*. Cambridge: Cambridge University Press.
Arrigoitia, M. (2014) Unmaking public housing towers: The role of lifts and stairs in the demolition of Puerto Rican project. *Home Cultures* 11 (2), 167–196.
Ashley, B., Hollows, J., Jones, S. and Taylor, B. (2004) *Food and cultural studies*. London: Routledge.
Bissell, D. (2008) Comfortable bodies: Sedentary affects. *Environment and Planning A* 40 (7), 1697–1712.
Blunt, A. and Dowling, R. (2006) *Home*. Abingdon: Routledge.
Blunt, A. and Varley, A. (2004) Introduction: Geographies of home. *Cultural Geographies* 11 (1), 3–6.
Brickell, K. (2012) 'Mapping' and 'doing' critical geographies of home. *Progress in Human Geography* 36 (2), 225–244.
Brickell, K. and Baxter, R. (2014) For home unmaking. *Home Cultures* 11 (2), 133–143.
Brickell, K. and Datta, A. (eds.) (2011) *Translocal geographies: Spaces, places and connections*. Aldershot: Ashgate.
Brun, C. and Lund, R. (2008) Making a home during crisis: Post-tsunami recovery in a context of war, Sri Lanka. *Singapore Journal of Tropical Geography* 29, 274–287.
Casey, E. (2009) *Remembering: A phenomenological study*. Bloomington: Indiana University Press.
Csikszentmihalyi, M. and Rochberg-Halton, E. (1981) *The meaning of things: Domestic symbols and the self*. Cambridge: Press Syndicate of the University of Cambridge.
Datta, A. (2008) Building differences: Material geographies of home(s) among Polish builders in London. *Transactions of the Institute of British Geographers* 33(4), 518–531.
Davidson, J. and Milligan, C. (2004) Editorial embodying emotion sensing space: Introducing emotional geographies. *Social and Cultural Geography* 5 (4), 523–532.
Easthope, H. (2004) A place called home. *Housing, Theory and Society* 21, 128–138.
Gorman-Murray, A. (2007) Reconciling self: Gay men and lesbians using domestic materiality for identity management. *Social and Cultural Geography* 9 (3), 283–301.
Gorman-Murray, A., McKinnon, S. and Dominey-Howes, D. (2014) Queer domicile: LGBT displacement and home loss in natural disaster impact, response, and recovery. *Home Cultures* 11(2), 237–261.

Guha-Sapir, D., Hoyois, P. and Below, R. (2013) *Annual disaster statistical review 2012: The numbers and trends.* CRED.

Guiliani, M. and Feldman, R. (1993) Place attachment in a developmental and cultural context. *Journal of Environmental Psychology* 13, 267–274.

Hawkins, R. and Mauer, K. (2011) 'You fix my community, you have fixed my life': The disruption and rebuilding of ontological security in New Orleans. *Disasters* 35 (1), 143–159.

Hooper-Greenhill, E. (2000) *Museums and the interpretation of visual culture.* London: Routledge.

Hyndman, J. (2007) The securitization of fear in post-tsunami Sri Lanka. *Annals of the Association of American Geographers* 97 (2), 361–372.

Kaika, M. (2004) Interrogating the geographies of the familiar: Domesticating nature and constructing the autonomy of the modern home. *International Journal of Urban and Regional Research* 28, 265–286.

Law, L. (2001) Home cooking: Filipino women and geographies of the sense in Hong Kong. *Cultural Geographies* 8 (3), 264–228.

Liu, L. (2014) 'A search for a place to call home: Negotiation of home, identity and sense of belonging among new migrants from the People's Republic of China (PRC) to New Zealand.' *Emotion, Space and Society* 10 (1), 18–26.

Longhurst, L. Johnston, L. and Ho, E. (2009) A visceral approach: Cooking at home with women in Hamilton, New Zealand. *Transactions of the Institute of British Geographers* 34 (3), 333–345.

Marcoux, J.S. (2001) The refurbishment of memory. In Miller, D. (ed.) *Home Possessions.* Oxford: Berg, 69–86.

Miller, D. (2010) *Stuff.* Cambridge: Polity Press.

Miller, D. (2008) *The comfort of things.* Cambridge: Polity Press.

Miller, D. (2001) Behind closed doors. In Miller, D. (ed.) *Home possessions: Material culture behind closed doors.* Oxford: Berg, 1–19.

Montasim, H. and Heynen, H. (2011) At home with displacement? Material culture as a site of resistance in Sudan. *Home Cultures* 8 (1), 43–70.

Moore, J. (2000) Placing home in context. *Journal of Environmental Psychology* 20 (3), 207–218.

Morrice, S. (2013) Heartache and Hurricane Katrina: Recognizing the influence of emotion in post-disaster return decisions. *Area* 45 (1), 33–39.

Pearce, S. (1992) *Museums, objects and collections: A cultural studies.* Washington, DC: Smithsonian Institution Press.

Petridou, E. (2001) The taste of home. In Miller, D. (ed.) *Home possessions: Material culture behind closed doors.* Oxford: Berg, 87–106.

Pile, S. (2011) For a geographical understanding of affect and emotion. *Transactions of the Institute of British Geographers* 36, 603–606.

Pile, S. (2010) Emotions and affect in recent human geography. *Transactions of the Institute of British Geographers* 35, 5–20.

Putnam, T. (1993) Beyond the modern home: Shifting the parameters of residence. In Bird, J., Curtis, B., Putnam, T., Robertson, G. and Tickner, L. (eds.) *Mapping the futures local cultures, global change.* London: Routledge, 150–165.

Rose, G. (2012) *Doing family photography: The domestic, the public and the politics of sentiment.* London: Ashgate.

Rose, G. (2003) 'Family photographs and domestic spacings: A case study.' *Transactions of the Institute of British Geographers* 28, 5–18.

Sherman, E. and Dacher, J. (2005) Cherished objects and the home: Their meaning and roles in late life. In Rowles, G. and Chaudhury, H. (eds.), *Home and identity in late life: International perspectives.* New York: Springer Publishing Company, 63–80.

Shumaker, S. and Conti, G. (1985) Understanding mobility in America: Conflicts between stability and change. *Home Environments* 8, 237–253.

Schroder, N. (2006) *Spaces and places in motion: Spatial concepts in contemporary American literature.* Tübingen: Gunter Narr Verlag.

Taylor, M. (2006) 'Furniture is a kind of dress': Interiors as projection of self. In McMinn, T., Stephens, J. and Basson, S. (eds.) *Contested terrains.* SAHANZ Conference proceedings. SAHANZ.

Tolia-Kelly, D. (2004) 'Locating processes of identification: Studying the Precipitates of re-memory through artefacts in the British-Asian home.' *Transactions for the Institute of British Geographers* 29, 314–329.

Tuan, Y-F. (1977) *Space and place: The perspectives of experience.* Minneapolis: University of Minnesota Press.

Tumarkin, M. (2005) *Traumascapes: the power and fate of places transformed by tragedy.* Melbourne: Melbourne University Press.

Ureta, S. (2007) Domesticating homes: Material transformation and decoration among low income families in Santiago, Chile. *Home Cultures* 4, 311–336.

van den Honert, R. and McAneny, J. (2011) The 2011 Brisbane floods: Causes, impacts and implications. *Water* 3, 1149–1173.

Vidler, A. (2002) *Warped space: Art, architecture and anxiety in modern culture.* Cambridge: MIT Press.

Vidler, A. (1992) *The architectural uncanny: Essays in the modern unhomely.* Cambridge: MIT Press.

Walsh, K. (2006) British expatriate belongings, mobile homes and transnational homing. *Home Cultures* 3 (2), 119–140.

Warde, A. (1997) *Consumption, food and taste.* London: Sage.

Watson, J. and Caldwell, M. (2005) *The cultural politics of food and eating: A reader.* Oxford: Blackwell.

Whitmore, H. (2001) Value that marketing cannot manufacture: Cherished possessions as links to identity and wisdom. *Generations* 25 (3), 57–63.

Wilford, J. (2008) Out of rubble: Natural disaster and the materiality of house. *Environment and planning D: Space and society* 26 (4), 642–662.

Wise, J. (2000) Home: Territory and Identity. *Cultural Studies* 14, 295–310.

Young, I. (1997) *Intersecting voices: Dilemmas of gender, political philosophy and policy.* Princeton: Princeton University Press.

5 'Goodnight, sleep tight'

Bedtime stories, picture-book bedrooms and tales of comfort

Jamie Adcock

> A bedroom scene is generative: from it springs a certain question-marked figure of the child, the designated offspring we are invited to read about, along with that child's story or fabula and the picture book itself, which, like a notepad… opening after opening, under the covers, reveals what we readers both fear and desire to *see* and to *know*, as proposed by the title of the book.
>
> (Moebius, 1991: 53, original emphases)

Not always navigated without tears, bedtime can be a time of stress and conflict between parents and children; for example, children often argue that they are not tired and may challenge their parents' insistence that it is time to go to bed. However, as others have observed, the bedtime story has long served as a means to coax children into bed and to relieve tension (Albert and Jones, 1977), and the picture book has been an important part of calming rituals and bedtime routines for generations (Spitz, 1999). Telling stories through their words and images, and encouraging the intimate and loving practice of child and parent reading together, picture books are intended to arouse and inspire the land of Nod and, as illustrated by the frequent depiction and use of bedroom scenes, are at the juncture between the waking world and the dream world (Moebius, 1991).

This chapter explores how picture books work as imaginative materials that prepare children for bed. Drawing upon a close reading of texts in which bedrooms feature as important settings or plot elements, it argues that such depictions establish some sort of connection or empathy with readers which facilitates children's going to sleep. Approaching the picture book as a text and object which is specifically read or used within the bedroom, it shows how the bedroom is a key device that situates both child and adult readers within the narratives of stories and how depicted bedrooms can synchronise with actual bedtime happenings. It considers how rooms on paper foster the creation of a comfortable and relaxed atmosphere conducive to sleep and how they might ease children's bedtime or sleep-related anxiety, or discomfort, such as that associated with the dark, the unknown of dreams, and monsters hiding under beds or in cupboards. It explores how picture books both present and perform bedrooms as spaces of safety, danger, fantasy and adventure and reveals the comforting messages embodied within. Although others have studied the picture-book genre, and some have analysed the same titles under examination here, the significance of the bedroom, when present within stories, is generally overlooked, and the space is often lost within readings that explore and identify other themes and ideas (for an exception, however, see Moebius, 1991). In focusing upon the bedroom, this chapter brings the space to the forefront of analysis and fully acknowledges its importance as a story element, or device.

Through an interpretative reading of two picture-book titles, this chapter explores the bedtime story as an everyday practice of home which mediates between homely comfort and unhomely discomfort. Divided into two analytical sections, it shows that bedtime stories not only reflect and aim to reproduce an idealised image of bedtime homeliness but that they also present bedtime as an uncomfortable and frightening experience for children. It argues that bedtime stories not only serve as parental tools to instil calm and comfort within children but that they also challenge children to step outside of their comfort zones and to face up to their anxieties in order to achieve comfort before going to sleep. The first section opens up the bedroom and bedtime scene depicted within Margaret Wise Brown's (1975 [1947]) classic bedtime story *Goodnight Moon* and examines how the tale reflects and feeds back into the experience of bedtime. With reference also to Russell Hoban's (1963) *Bedtime for Frances*, it considers how *Goodnight Moon* settles and relaxes children before sleep, how children are able to relate themselves to the story, and how the book attempts to alleviate their concerns about going to sleep. With a particular focus upon Mercer Mayer's (1976) *There's a Nightmare in My Cupboard*, the second section explores children's nighttime anxieties and the terror of monsters and unknown life forms lurking within dark corners and spaces of bedrooms. It reveals how the story motivates children to confront and engage with their fears at bedtime and how it encourages parents to be supportive and understanding in such situations. The chapter concludes by reflecting upon the role of bedrooms within bedtime stories, and the lessons, or wisdom, that they impart. First, however, it provides a brief introduction to both picture books and the bedtime story.

Picture books: an introduction

Usually, but not always, aimed at a younger audience, picture books are a unique form of literary text. As Nodelman (1988: vii) outlines, they 'communicate information or tell stories through a series of many pictures combined with relatively slight texts or no texts at all'. A distinctive feature of the genre, pictures generally fill the page and hold and transmit most of the detail of a story, as the text, if there is any, is often brief and concise

(Nodelman, 1988). The 'essence of the picture book', however, lies in the relationship between the words and the pictures, and both are required for a narrative to progress and develop (Sipe, 1998: 97). Nodelman (1988: viii) observes that 'the words and the pictures in picture books both define and amplify each other' and that their interaction contributes to 'unique rhythms, unique conventions of shape and structure, [and] a unique body of narrative techniques'. Indeed, Hunt (2001: 288, original emphasis) argues that to read a picture book, 'we have to adjust our *pace* and manner of reading and to relate non-linear "reading" of the picture to linear processing of words'. For example, whilst pictures may illustrate the meaning of words and fill in on any missing detail (Nodelman, 1988; Sipe, 1998), he explains that they sometimes contradict the text and create confusion or ignore it and present an entirely different story. Moreover, and reflecting the innocence and young age of their intended readers, picture books are frequently noted for their educational benefits and for teaching children about the ways of the world and society (Nodelman, 2004). As Nodelman (2004: 157) notes, picture books illustrate 'what the world implied by the words looks like', and children learn about 'the world outside the book in terms of the images within it'.

Time for bed: the bedtime story

> Children... need rooms of their own, places – whether real or imaginary – of peace and well being and unconditional love, places where a secure sense of self can begin to grow. That solid sense of being at home in the world is one of the great gifts the adult world has the power to bestow on its children.
>
> (Marcus, 1997: n.p., cited in Galbraith, 1998: 176)

A defining feature of the picture book is 'the drama of the turning of the page' (Bader, 1976: 1, cited in Moebius, 1991: 53). Works of textual and pictorial art (Nodelman, 1988; Stanton, 1990), picture books stimulate the imagination and provoke a sense of wonder. A text that 'tends to play to an indoor audience' and 'which favors darkening rooms', the picture book is often a precursor to sleep (Moebius, 1991: 53). Used nightly at bedtime to settle children and to ease them into the realm of dreams, it underlines the intimacy and tenderness of the parent–child bond. A material object that parents and children can clasp, touch and share together (Albert and Jones, 1977; Stanton, 1990), the picture book, as Spitz (1999: 2) observes, calls for 'the practice of reading aloud to young children'. She argues that picture books 'require the participation of warm, breathing adult human partners who have available laps, keen eyes and ears, arms adept at holding while turning pages, and perhaps a flair for the dramatic' (Spitz, 1999: 2). Nestled together and partners in reading, adult narrator and child listener share in the drama and excitement of a story and 'are united in experiencing its pains and pleasures' (Albert and Jones, 1977: 114). Interactive and relational, a child may look intently at the pictures and follow the text whilst listening to the spoken narrative, and the adult may answer questions and act as a kind of informational repository (Albert and Jones, 1977; Spitz, 1999). Together, words and pictures envelop children within realms of fantasy and enchantment which help to create restful atmospheres of calm and peace. A means to avoid and alleviate possible bedtime arguments and protests, the bedtime story enthrals and captivates and, as Albert and Jones (1977: 115) note, allows children 'to let go of the reality of the setting in which the storytelling is taking place'. A device intended to lull children to sleep, the picture book thus facilitates the transition from daytime activity to night-time rest, and this elemental function of the genre is often reflected within the symbolic representation of bedrooms and bedtime happenings within stories (Moebius, 1991).

The bedtime scene: *Goodnight Moon*

> How many little children, faces freshly scrubbed, snuggly in pile pajamas, have gone sleepily into [this] enchanted evening? How many mothers and fathers have led them there, just as entranced as the children? Entranced by the vivid solidity of the sleepy rabbit's room, the reassuring roster of the child's everyday tools: the comb, the brush, the mittens. And by [the] playful mouse, here on the windowsill, there on the hearth.
>
> (New York Times, 1988)

A delightful 'goodnight, sleep-tight' tale, *Goodnight Moon* is perhaps the archetypal bedtime story. Written by Margaret Wise Brown and brought to life visually through the enchanting illustrations of Clement Hurd, it depicts a nightly bedtime ritual in which an anthropomorphic child rabbit bids 'goodnight' to his bedroom, 'the great green room' (Brown, 1975 [1947]: 1), and to various objects both within and outside the space. Described by Marcus (1987) as an 'unconventional design' of book, the illustrations repeatedly switch from colour panoramic landscapes of the 'great green room' to single black-and-white images of saluted objects located within the bedroom or seen through the windows (Galbraith, 1998). A poem structured by 'auditory and pictorial patterns of flow and forward movement' (Spitz, 1999: 28), the story draws sleepy children into its world and perfectly captures the wonderful innocence of a young child's reality (Marcus, 1987). Within the 'great green room', as Marcus (1987) reflects, 'the far-flung moon and stars, the fantastic nonsense of moon-jumping cows and the mysterious "nobody," to whom goodnight is said, take their places beside mundane clocks and socks in an ample, fiercely resilient vision of the very young's Here and Now world'. An inventory, or catalogue, of items which are identified and metaphorically 'put to bed by the bunny child', the book encourages children to spot the things that the rabbit highlights, both on the page and in their own bedrooms where the reading is taking place (Nodelman, 1988; Susina, 1998: 118). However, there are many features within the 'great green room' that the text fails to mention (Galbraith, 1998). As Nodelman (1988: 210) observes, 'The delight viewers feel in discovering these things with their own eyes rather than their ears reveals how basic and important is the difference between the information available in words and in pictures.' For example, a little mouse moves around the room from page to page and 'provides a game of hide and seek for the reader' (Stanton, 1990; Galbraith, 1998: 176). Furthermore, a framed picture hanging on the left-hand wall depicts an illustration from another of Brown and Hurd's picture books, *The Runaway Bunny* (1942), and there is a copy of the book on a bookshelf (Nodelman, 1988; Galbraith, 1998; Susina, 1998). Children may also notice that a book on the bedside cabinet is an edition of *Goodnight Moon*. A self-referential and relational object (Galbraith, 1998), this small detail allows children to read themselves into the story and to feel a sense of connection with the rabbit child on the page.

The 'great green room' is cosy and inviting. A roaring open fire, fluffy kittens and an old lady rabbit knitting whilst sitting in a rocking chair form part of a magical bedroom world at bedtime (*New York Times*, 1988). Tucked into bed, the central character, Bunny – the child's name revealed by a labelled hairbrush – has presumably just been put to bed; the closed copy of *Goodnight Moon* on the bedside table is a sign that he has been read to, perhaps by the old lady, and that it is now time to settle down and to go to sleep. Indeed, Bunny begins to bid his room, the objects within it and the outside world 'goodnight'. The little rabbit's poetic naming of his bedroom environment, constituting the text of the story, serves as a 'farewell to the world of day' and marks the beginning of his onward journey into the realm of sleep (Susina, 1998: 119). As Bunny's 'goodnight' ritual progresses, the 'great green room' becomes ever darker and the ambient light gradually fades (Stanton, 1990), thus signalling that the time for sleep is slowly approaching for both the rabbit child and children reading or listening to the story within the real world. Contributing to the seemingly perfect bedtime scene, the old lady grandmother figure remains in her chair until Bunny falls asleep, whilst the kittens play with her wool. The final image of Bunny sound asleep depicts a world at peace. 'Goodnight noises everywhere' (Brown, 1975 [1947]: 30) pervade the bedroom, the kittens are curled up and sleeping on the old lady's chair, and the intensity of the full moon and stars shining through the windows hint at the magic and wonder of the dream world to which Bunny has departed.

Observing that picture books are often read as part of bedtime rituals, Moebius (1991: 55) argues that 'it is not surprising that bedroom scenes, when they do not appear throughout, should dominate their final pages'. He claims that the depiction of the bedroom at or towards the end of a story signifies that the tale will soon finish and 'will have nothing more to say'

(Moebius, 1991: 55). As he explains, 'No matter what the book is about, the bedroom betokens the restoration of calm and the absence of confusion or anxiety. The bedroom scene is the book's (and the sleepy child's) destiny, its calling' (Moebius, 1991: 55). An 'illustration of synchronous arrival' (Albert and Jones, 1977: 129), therefore, the symbolic appearance of the bedroom works to bring exciting adventures and wonderful journeys shared by both child protagonists and child readers to a close. It indicates that the end of the cosy parent-child reading encounter is imminent, that the book will soon be returned to the bookshelf and that the time to turn out the light is almost nigh. The 'independent reality' of the picture book essentially draws adult narrators and child listeners into the chronological sequence of its narrative (Albert and Jones, 1977: 114; Moebius, 1991). For example, tucked into bed and following the story of *Goodnight Moon*, perhaps with a favourite teddy bear and with heavy eyes, children are able to relate to the rabbit, who is also in bed, and understand that their own bedtimes are coming. The final image of Bunny sound asleep closes the story and is a perfect illustration that it is time for child readers to go to sleep as well.

As parents and guardians of small children may testify, bedtime can be an occasion fraught with tension and disagreement. As Albert and Jones (1977: 114) highlight,

> Putting the child to sleep is an exercise in the rights and the treacheries of authority. The adult has the power to leave the child, to have him [*sic*] exiled to his proper place, and the child has the will to protest with all the resources at his command.

Goodnight Moon, however, aims to produce a sense of homely comfort and to calm and appease children. It presents and reflects a particular image, or imaginary, of home, a domestic ideal of warmth, hearth, family and intimacy, and attempts to reproduce this image through its reading. Indeed, the 'great green room' is a peaceful, tranquil and 'eloquent expression of at-homeness' (Marcus, 1987), an idyllic bedtime scene which child readers are able to participate and share in. Moreover, the story serves as a verse-like lullaby, or 'hypnotic bedtime litany' (Prager, 2000: C3, cited in Robertson, 2000: 206), which children are able to recite and repeat before bed themselves. For example, Spitz (1999: 29) argues that the words, when spoken aloud, provide 'an auditory counterpart and complement for a child's heartbeat at it calms down in the moments before' sleep; she notes that the text is characterised by 'rhythmic structures' and that the book soothes through alliteration, rhyme and 'the regular beats of the accented words'. She also suggests that the book serves as a kind of 'antidote' to the fast pace of modern everyday life and allows children to 'relax into it', reasoning that the book 'refuses speed' and 'cannot be hurried through' (Spitz, 1999: 30).

The reading of *Goodnight Moon*, and the nightly routine of bedtime stories more generally, illustrates the meaning-making processes through which homes are made and remade and the material and emotional relationships upon which the home is founded. The book underlines that the home 'does not simply exist' but that it is 'continually created and recreated through everyday practices' which have 'both material and imaginative elements' (Blunt and Dowling, 2006: 23; also see Miller, 1988; Rose, 2003, 2004; Hurdley, 2006). Moreover, it illustrates that the home is 'a field of rules' which requires children to act, or behave, in certain ways (Wood and Beck, 1994: 1). For example, at no point does the rabbit child within the story question or dispute that it is time for bed. Reflecting that bedtime is an adult-defined practice in which children must turn out the light and go to sleep when instructed, and adhering to the 'rules' of going to bed, the protagonist seems to accept his bedtime without any objections and is an epitome of good behaviour for child readers to emulate. *Goodnight Moon* therefore illustrates how picture books work to teach children about the ideologies of culture and society (Nodelman, 2004). Often encountered at a young age (Nodelman, 1988), they not only encourage and contribute to the development of children's language and communication skills (Stephens, 1989), but they also enhance readers' knowledge about life, the world and their place within it (Nodelman, 2004). As Nodelman (2004: 159) observes, 'Like most narrative, picture-book stories most forcefully guide readers into culturally acceptable ideas about who they are through the privileging of [a particular] point of view'. For example, he argues that a theme of many picture books is the 'need for adult authority' and the 'inevitability of child-like irresponsibility' (Nodelman, 2004: 159).

It is also important to acknowledge that the child protagonist within *Goodnight Moon* is a rabbit. The book belongs to a category of literature almost unique to children's stories in which animal characters act and live like humans (Hunt, 1994). Tales which feature 'humanized animals' (Nodelman, 1988: 35), according to Flynn (2004: 420), can be traced to the long-held belief that children are able to relate to such characters and that they have some sort of 'innate sympathy or connection with animals'. Embodied within social and cultural constructions of childhood, and derived from late-eighteenth- and early-nineteenth-century Romantic thinking, an animistic relationship is said to exist between animals and children (Flynn, 2004). Furthermore, whilst adults are thought to be 'divorced from nature', children are often considered to be closer to the natural world and ignorant of certain human-defining 'markers and rules' (Cosslett, 2002: 476, cited in Flynn, 2004: 420). As Nodelman (2004: 161) notes, the depiction of human-like animals within picture books is 'a metaphor for the state of human childhood, in which children must learn to negotiate between the animal-like urges of their bodily desires and the demands of adults that they repress desire and behave in socially acceptable ways'. Indeed, the sweet little bunny within *Goodnight Moon* is a paragon of good manners and reflects the practice of associating animal instinct and perceived animal behaviour with human conduct (Nodelman, 1988; Flynn, 2004). For example, within many picture-book stories, and throughout culture and society more generally, foxes are frequently seen as cunning, pigs are depicted as greedy, and rats are spiteful (Nodelman, 1988). Moreover, Nodelman (1988: 116) suggests that furry little animals are recurring picture-book characters because they are also 'small enough to express the traumas of small children in a world of large adults'. In essence, they universalise and enable all children to empathise with them (Hunt, 1994; Spitz, 1999). Probably not a concern at the time of writing or publication, Bunny, for example, has no discernible race but, perhaps more significantly, is also age and gender neutral; though the blue-and-white stripy pyjamas that Bunny is wearing imply that the little rabbit is male, the child's sex is largely undefined (Galbraith, 1998; Robertson, 2000). Bunny therefore appeals to both boys and girls of all ages; this is particularly noteworthy, as few picture books are formed around central female characters, and girls often have to relate to male protagonists and role models (Spitz, 1999).

Goodnight Moon also touches upon children's concerns about going to sleep. Robertson (2000: 204), for example, argues that 'the story challenges the child listener to enter into a relationship with the text' and 'to engage with and work through the perils of bedtime'. As Spitz (1999) observes, the book works to relieve sleep-related anxiety through the repetition and predictability of its words and images and through the restful atmosphere that its reading generates. It aims to soothe and relax children in the moments leading up to sleep (Spitz, 1999) and at the very point that they are required to shut their eyes and enter the unknown and unfamiliar realm of the dream.

According to Gaddini (1996: 28), 'Falling asleep, in psychological terms, means losing control of reality and plunging into a world of fantasy.' Reflecting upon the infant fear of sleep, she argues that young children are generally unable to distinguish between 'external and internal realities' (Gaddini, 1996: 28). As she observes,

> No anxiety can be compared in intensity and duration to the anguish experienced by an infant when he [*sic*] shuts his eyes and consequently loses control of external reality, which is objective, shared and constant, and, to some extent, predictable.
>
> (Gaddini, 1996: 28, cited in Robertson, 2000: 204)

Spitz (1999) similarly claims that a child's sense of who they are is relatively fragile at an early age and suggests that young children may associate the unconsciousness of sleep with the loss of individual identity. As Albert and Jones (1977: 116) also note, young children are yet to fully understand 'the permanence of the social world' and 'the fact that events unfold in time even when we are asleep'. Alluding to the parallels between sleep and death (Spitz, 1999; Robertson, 2000), they explain that 'while [a child] may fear that he [*sic*] may miss something, the greater fear is that the world will stop when he goes to sleep, that there will be nothing coming next' (Albert and

Jones, 1977: 116). Bunny's ritual of saying 'goodnight' to the things within the 'great green room' thus holds a particular significance. An expression that points to the taking of leave, 'goodnight' not only refers to imminent sleep but may also denote a departure, or a goodbye (Galbraith, 1998). As Galbraith (1998: 177) highlights, for example, should Bunny's penultimate 'goodnight' to the air be read and understood 'as a farewell rather than as an acknowledgement, then the ending of *Goodnight Moon* implies the death of the bunny in the bed'.

From a psychological perspective, therefore, protests before bed and efforts to delay sleep could be interpreted as survival strategies through which children attempt to hold on to 'a reassuring external reality' at bedtime (Gaddini, 1996: 28, cited in Robertson, 2000: 205). A child's endeavours to put off sleep and to stay awake at bedtime are perfectly illustrated by Russell Hoban's (1963) *Bedtime for Frances.* Within this classic picture-book narrative, a little girl badger asks her parents for a glass of milk, a piggyback and kisses before going to bed. Having been tucked into bed by her parents, she then continually gets out of bed to tell them that there is a tiger and a giant in her room, that something is moving her bedroom curtains and that she has forgotten to brush her teeth. In her analysis of the story, Spitz (1999) notes that the title page of the book depicts a nervous-looking Frances, unsure of her parents' reaction, peering out from behind a door after having got out of bed. Picturing one of 'those paradigm moments in childhood', the illustration, she argues, not only encapsulates the little girl's hopeful optimism that her parents will allow her to stay up with them but also conveys her efforts to preserve a bond and loving structure that defines her existence (Spitz, 1999: 44). It also symbolises children's need for affection and dependency, and an aspect of human development as summarised by attachment theory, a concept which maintains that the parent-child relationship 'forms the basis of each child's (and therefore adult's) core sense of being real, here, and alive' (Galbraith, 1998: 172; for example, see Holmes, 2014). With this in mind, Galbraith (1998: 175) suggests that in the absence of his parents, Bunny transfers his feelings of attachment 'from person to environment' within *Goodnight Moon.* She claims that as Bunny bids 'goodnight' to his room, he addresses the objects and imaginative entities within as if they are people. Perhaps an attempt to delay sleep like Frances – an hour and ten minutes pass from the first to the last page, as indicated by the clocks on the fireplace and the bedside cabinet – Bunny's naming of his bedroom surroundings works, as Spitz (1999: 31) observes, to reinforce his 'sense of intactness just at the moment when that cohesion seems to be slipping away'.

The bunny child's 'map of existence' also points to the idea that individuals are constituted in relation to objects (Robertson, 2000: 210). Indeed, the items that Bunny bids 'goodnight' and those on display within the 'great green room' speak to the practice and rituals of everyday life and serve as 'reassurances of the ongoing nature of the self' (Spitz, 1999: 32). For example, a brush and comb are specifically set out on a table for the next morning, mittens and socks are drying in front of the fire ready for use tomorrow, stuffed animals on the bookshelf await future play and the books hint at stories still to be read, with the open copy of *The Runaway Bunny* perhaps a clue to tomorrow night's bedtime story. Furthermore, as Spitz (1999: 34) observes, the objects within the room remain visible throughout the story despite the ever-increasing darkness, and, although they may not be as clear in the gloom, the prominence of their outline teaches children about 'object permanence' and that 'life has stability, reliability, and durability'. Moreover, the consistent appearance of the objects and Bunny's repeated naming of them are means to ease children's fear of the dark by undermining its power to conceal and cover. As Spitz (1999: 34) explains, the book emphasises that 'just because something cannot be seen at a particular moment it has not disappeared forever. One's loved ones are a permanent acquisition, even when lost to view.' Soothing at the onset of darkness, *Goodnight Moon* therefore affords children the confidence to close their eyes and to relinquish their hold on reality, smoothing their transition into the world of dreams.

Nightmares before bed: *There's a Nightmare in My Cupboard*

Although it is not clear whether the little rabbit within *Goodnight Moon* is afraid of the dark, it is a fear that many children possess and can provoke, as Galbraith (1998: 173) observes, an 'increased reliance on touch and close proximity' at bedtime. Darkness is often a source of much anxiety for children, both young and old, and can arouse concerns that might otherwise remain hidden during daylight hours (Galbraith, 1998). Shrouded in gloom and obscured by dim light, darkened bedrooms are potentially intimidating and frightening for young children, particularly at bedtime. As Galbraith (1998: 173) explains, 'Without the reassurance of one's attachment figure, sounds and shadows that one would not ordinarily notice loom strange and fearsome.' Further reflecting Gaddini's (1996: 28) argument that they are not always able to differentiate between 'external and internal realities' (cited in Robertson, 2000: 204), young children may sometimes sense a sinister presence lurking within the dark and fear that something is waiting to 'get' them. For example, and as shown within *Bedtime for Frances* above, monsters, ghosts and dangerous animals are often imagined hiding under beds or inside wardrobes, cupboards and dark corners.

Children's bedtime discomfort and night-time anxiety is a central theme of many picture books. Through the depiction of uncomfortable bedtime experiences, these stories, when read at bedtime, often encourage children to face up to their fears and to work through their anxieties. Exemplified by Mercer Mayer's (1976) timeless *There's a Nightmare in My Cupboard*, many picture books challenge child readers to step outside of their comfort zones at bedtime in order to find inner comfort. First published in the United States as *There's a Nightmare in My Closet*, this well-known picture book, narrated in the first person from the unnamed protagonist's perspective

(Spitz, 1999), describes how a little boy is afraid of a 'nightmare', or monster, that lives inside the dark of his bedroom cupboard and how he successfully triumphs over his fear.

Upon opening the book, young readers are presented with a bedtime scene with which they may identify. Illuminated by the faint light of a bedside lamp, the boy is shown lying in bed and staring nervously at the cupboard, the door of which is slightly ajar but hiding whatever is lurking within. Frozen with fear, he pulls his blanket up tight to his face and cocoons himself as the night shadows become longer around him. Perhaps contributing to his anguish and increasing the feeling of terror, a strong wind appears to be blowing through an open window, for his bedroom curtains have taken on a life of their own; the brilliance of the white frills stretched out and set against the dark green background of the bedroom walls transform them into ghostly apparitions. After much hesitation, the boy musters the courage to leave the protective environment of his bed and goes to shut the cupboard door, a significant act of bravery which, readers are informed, he performs every night at bedtime before going to sleep. However, unable to lock the door – despite the presence of a lock, there is no key – he arms himself with a cork popgun and toy cannon and takes them to bed in case the nightmare should escape into his bedroom, comforting objects and defensive weapons that illustrate just how frightened he is and which point to the mortal danger that he senses.

Exploring the source of children's night-time fears and how they can be treated, or resolved, from a behavioural perspective, King *et al.* (1997: 441) argue that night-time anxieties are 'experienced by nearly all children during the normal course of development'. They observe that such fear is very real to children and able to 'cause considerable personal distress' but that it can often be mistaken by parents for bad behaviour (King *et al.*, 1997: 441). For example, sometimes severe enough to hamper the running of everyday life, night-time fear can provoke a variety of child coping strategies which can lead to delayed and highly charged bedtimes (King *et al.*, 1997; Muris *et al.*, 2001). Frequently seen by parents and guardians as disruptive and irritating behaviour (Galbraith, 1998), children's coping strategies at bedtime, besides cuddling a teddy bear or other inanimate objects, include calling out to parents, repeatedly getting out of bed, attempts to stay up or to sleep with parents or siblings, and leaving televisions, radios and lights switched on (Graziano *et al.*, 1979, cited in King *et al.*, 1997: 432; also see Mooney *et al.*, 1985; Muris *et al.*, 2001). However, perhaps 'propaganda for a parental agenda' (Galbraith, 1998: 174), and unlike Frances when she perceives a tiger and a giant hiding within the dark of her bedroom, the little boy within *There's a Nightmare in My Cupboard* does not cry out for his parents or attempt to delay his bedtime. Reflecting a Western belief that has long held that children should sleep away from their parents and be able to soothe themselves when upset (Galbraith, 1998), the boy decides to confront his fear on his own. The epitome of a brave little soldier, he recruits some toy soldiers and puts on an army helmet for protection, crouches down behind his pillow, turns out the light and aims his cannon and popgun towards the cupboard in anticipation that the monster will reveal itself. When he hears the nightmare creep out of the cupboard, he surprises it by turning on the light, orders it to 'go away' (Mayer, 1976: 14) and threatens to shoot it.

Liminal in form, monsters within literature are often imagined and depicted as having bodily features that belong to a variety of different animal species (Nuzum, 2004). Drawing upon Cohen's (1996: 6–7) interpretation of the meaning of the word 'monster', Nuzum (2004: 208) argues that monsters 'are neither one thing nor the other' and suggests that their hybrid nature contributes to their scariness. A stereotypical giant green monster, the boy's cupboard nightmare is similarly a cross-breed of various animals; for example, it has the big pointy ears of an elephant, a bushy mane like a horse, stubby legs and short arms akin to a human, and a dragon-like tail. Significantly, however, it is also rather childlike and, as Spitz (1999) notes, quite goofy in appearance. Furthermore, it looks like a friendly creature and does not seem to pose a threat to the boy; it only has two very small front teeth, which are noticeably blunt rather than sharp, and its fingers and toes have no visible claws. Reassuring children reading the story that nightmares, both before and during sleep, are actually far from scary and that bedroom monsters and other imaginary creatures cannot hurt them, the nightmare is clearly more harmless than harmful and appears to be nervous and shaking with fear itself. When the boy shoots at it, for instance, it bursts into tears and acts like an oversized baby or toddler (Spitz, 1999). Such behaviour is typical of many picture-book bedtime monsters, for they are often depicted as cowardly and defenceless creatures who are scared of children, reinforcing the idea that they are more frightened of children than children are of them.

The ironic depiction of the huge nightmare crying its eyes out like a big baby perhaps encourages child readers to reflect upon their own behaviour when distressed at bedtime; for example, the ridiculousness and silliness of such a large creature in tears invites the scorn and derision of small children (Spitz, 1999). As the story progresses, however, it becomes apparent that the monster is frightened of something within the cupboard, that it does not want to sleep there and that it has tiptoed into the boy's bedroom looking for solace and comfort. A reflection of a child's night-time coping strategy, though one where the roles of the protagonists are reversed and re-imagined, the nightmare does not wish to harm or haunt the boy but is actually a child itself, a little monster, seeking moral support and reassurance from a wise and strong parent-like figure. Signalling parental frustration and impatience at a child's tears (Spitz, 1999), the little boy is angry and frustrated at first and offers the nightmare little relief by telling it to 'be quiet' (Mayer, 1976: 19). As Spitz (1999) suggests, the boy's initial lack of empathy for the monster child's plight prompts both child and adult readers to feel sorry for the nightmare, sympathy which reinforces the idea that children can only overcome their fears and bedtime anxieties with a little parental understanding.

A display that serves to undermine the notion that children's desire to remain close to their parents and cries for assistance at bedtime are 'undesirable habits to be extinguished' (Galbraith, 1998: 173), the little boy eventually softens his stance and invites the nightmare into his bed for comfort (Spitz, 1999). Effectively providing the monster with 'the parenting he is missing' himself (Galbraith, 1998: 178), he illustrates to parents and guardians reading the story that they should always acknowledge their children's fears, take them seriously and never dismiss or ignore them. Emphasising that there is nothing wrong with being afraid, though – for example, King *et al.* (1997) suggest that children often feel embarrassed and ashamed in such situations – the boy's support for the nightmare also informs child readers that they should never have to suffer alone or in silence. Indeed, the book's penultimate image of the boy and the nightmare smiling in bed together highlights how a little parental support can make everything better, and demonstrates that children should never hesitate to tell their parents whenever they are feeling anxious or unhappy. It also speaks to the importance of children confronting their fears and to the pride and relief that they should feel in doing so; the little boy, for example, remarks, 'I suppose there's another nightmare in my cupboard, but my bed's not big enough for three' (Mayer, 1976: 26). This is reinforced within the final wordless illustration, for a second monster emerges from behind the cupboard door when the boy and nightmare are finally asleep to illustrate that 'fears, even those that have been dealt with successfully, tend to recur' (Spitz, 1999: 67). However, given that the boy and monster are depicted in bed sleeping peacefully alongside one another, the image further cements the idea that with a little bravery, comfort and mutual understanding, children and parents can always triumph over such fears together.

Conclusion

Although *Goodnight Moon* and *There's a Nightmare in My Cupboard* follow different narratives and characters and use and present bedrooms in different ways in support of their plots, both have something significant in common. In keeping with the educational potential of picture books and the young age of their intended audience (Nodelman, 2004), each provides children with a lasting lesson and enhances knowledge. Whilst their reading creates momentary pleasure, each book bestows upon readers a kind of 'psychological gift' (Albert and Jones, 1977: 115). Because each child possesses a bedroom, the equivalent space on the page is a key device, or medium, for such teaching and enables the books to establish some sort of connection with readers. By featuring and depicting the bedroom in some shape or form, the stories are able to construct plotlines and scenes and show situations and protagonists with which child readers can identify. For example, when read or used within the bedroom, they are able to synchronise themselves to, and may feed back into, real-life bedroom happenings. At bedtime, for instance, their words and pictures are intended to facilitate children's sleep and the development of dream worlds. However, their contrasting storylines illustrate how bedtime stories mediate between the themes of comfort and discomfort. *Goodnight Moon*, for example, aims to produce the perfect conditions for sleep through its depiction of a seemingly perfect and homely bedtime scene. Demonstrating that the reading of bedtime stories is an everyday practice of homeliness, it seeks to calm and comfort children and to allay their sleep-related anxieties by generating a cosy and relaxed atmosphere and recreating the idealised image of bedtime depicted. In contrast, *There's a Nightmare in My Cupboard* portrays bedtime as a potentially frightening and uncomfortable experience for children. It encourages young readers to test their limits and to explore outside of their comfort zones in order to overcome discomfort and invites adults to reflect upon their attitudes and reactions to children's night-time fears. Both highlight the significance of bedtime reading as a means to shore up children's confidence and self-esteem, at bedtime as well as more generally, and confirm that comfort at bedtime is as much about feeling at ease in one's surroundings as it is about being tucked up in bed.

References

Albert, S., Jones, W. (1977) 'The Temporal Transition from being Together to being Alone: The Significance and Structure of Children's Bedtime Stories', in Gorman, B.S., Wessman, A.E. (eds.) *The Personal Experience of Time*, New York: Plenum Press, pp. 111–132.

Bader, B. (1976) *American Picturebooks from Noah's Ark to the Beast Within*, New York: Macmillan.

Blunt, A., Dowling, R. (2006) *Home*, London: Routledge.

Brown, M.W. (Hurd, C.) (1975) [1947] *Goodnight Moon*, Tadworth: World's Work.

Cohen, J.J. (ed.) (1996) *Monster Theory: Reading Culture*, Minneapolis: University of Minnesota Press.

Cosslett, T. (2002) 'Child's Place in Nature: Talking Animals in Victorian Children's Fiction', *Nineteenth-Century Contexts*, 23 (4): 475–495.

Flynn, S. (2004) 'Animal Stories', in Hunt, P. (ed.) *International Companion Encyclopedia of Children's Literature*, 2nd edn, Vol. 1, London: Routledge, pp. 418–435.

Gaddini, R.B. (1996) 'Lullabies and Rhymes in the Emotional Life of Children and No Longer Children', *Winnicott Studies*, 11: 28–41.

Galbraith, M. (1998) '"Goodnight Nobody" Revisited: Using an Attachment Perspective to Study Picture Books about Bedtime', *Children's Literature Association Quarterly*, 23 (4): 172–180.

Graziano, A.M., Mooney, K.C., Huber, C., Ignasiak, D. (1979) 'Self-Control Instructions for Children's Fear Reduction', *Journal of Behavior Therapy and Experimental Psychiatry*, 10 (3): 221–227.

Hoban, R. (Williams, G.) (1963) *Bedtime for Frances*, London: Faber and Faber.

Holmes, J. (2014) *John Bowlby and Attachment Theory*, 2nd edn, London: Routledge.

Hunt, P. (1994) *An Introduction to Children's Literature*, Oxford: Oxford University Press.

Hunt, P. (2001) *Children's Literature*, Oxford: Blackwell.

Hurdley, R. (2006) 'Dismantling Mantelpieces: Narrating Identities and Materializing Culture in the Home', *Sociology*, 40 (4): 717–733.

King, N., Ollendick, T.H., Tonge, B.J. (1997) 'Children's Nighttime Fears', *Clinical Psychology Review*, 17 (4): 431–443.

Marcus, L. (1987) 'A Moon That Never Sets', *New York Times*, 25th January 1987. Available at: http://www.nytimes.com/1987/01/25/books/children-s-books-a-moon-that-never-sets.html?pagewanted=2&src=pm (accessed 8th August 2010).

Marcus, L.S. (1997) 'Commentary', in Brown, M.W. (ed.) *Goodnight Moon: A 50th Anniversary Retrospective*, New York: HarperCollins, n.p.

Mayer, M. (1976) *There's a Nightmare in My Cupboard*, London: J.M. Dent and Sons.

Miller, D. (1988) 'Appropriating the State on the Council Estate', *Man*, 23 (2): 353–372.

Moebius, W. (1991) 'Room with a View: Bedroom Scenes in Picture Books', *Children's Literature*, 19: 53–74.

Mooney, K.C., Graziano, A.M., Katz, J.N. (1985) 'A Factor Analytic Investigation of Children's Nighttime Fear and Coping Responses', *Journal of Genetic Psychology*, 146 (2): 205–215.

Muris, P., Merckelbach, H., Ollendick, T.H., King, N.J., Bogie, N. (2001) 'Children's Nighttime Fears: Parent-Child Ratings of Frequency, Content, Origins, Coping Behaviors and Severity', *Behaviour Research and Therapy*, 39 (1): 13–28.

New York Times (1988) 'Topics of the Times: Tools of Childhood', 14th February 1988. Available at: http://www.nytimes.com/1988/02/14/opinion/topics-of-the-times-tools-of-childhood.html?scp=2&sq=Clement+Hurd&st=nyt (accessed 6th August 2010).

Nodelman, P. (1988) *Words about Pictures: The Narrative Art of Children's Picture Books*, Athens, GA: University of Georgia Press.

Nodelman, P. (2004) 'Picture Books and Illustration', in Hunt, P. (ed.) *International Companion Encyclopedia of Children's Literature*, 2nd edn, Vol. 1, London: Routledge, pp. 154–165.

Nuzum, K.A. (2004) 'The Monster's Sacrifice – Historic Time: The Uses of Mythic and Liminal Time in Monster Literature', *Children's Literature Association Quarterly*, 29 (3): 207–227.

Prager, J.H. (2000) 'Moon Child', *The Citizen*, 8th October, pp. C2–C3.

Robertson, J.P. (2000) 'Sleeplessness in the Great Green Room: Getting Way under the Covers with *Goodnight Moon*', *Children's Literature Association Quarterly*, 25 (4): 203–213.

Rose, G. (2003) 'Family Photographs and Domestic Spacings: A Case Study', *Transactions of the Institute of British Geographers*, 28 (1): 5–18.

Rose, G. (2004) '"Everyone's Cuddled Up and It Just Looks Really Nice": An Emotional Geography of Some Mums and Their Family Photos', *Social and Cultural Geography*, 5 (4): 549–564.

Sipe, L.R. (1998) 'How Picture Books Work: A Semiotically Framed Theory of Text-Picture Relationships', *Children's Literature in Education*', 29 (2): 97–108.

Spitz, E.H. (1999) *Inside Picture Books*, New Haven, CT: Yale University Press.

Stanton, J. (1990) '"Goodnight Nobody": Comfort and the Vast Dark in the Picture-Poems of Margaret Wise Brown and Her Collaborators', *Lion and the Unicorn*, 14 (2): 66–76.
Stephens, J. (1989) 'Language, Discourse, and Picture Books', *Children's Literature Association Quarterly*, 14 (3): 106–110.
Susina, J. (1998) 'Children's Reading, Repetition, and Rereading: Gertrude Stein, Margaret Wise Brown, and *Goodnight Moon*', in Galef, D. (ed.) *Second Thoughts: A Focus on Rereading*, Detroit, MI: Wayne State University Press, pp. 115–125.
Wood, D., Beck, R.J. (1994) *Home Rules*, Baltimore, MD: Johns Hopkins University Press.

Section two

Difference and encounter

6 The geopolitics of (dis)comfort and indifference in Israel-Palestine

Daniel Webb

Introduction

> I am riding the train back home from El Sal to the Damascus Gate after visiting friends in Shuafat. It is swelteringly hot outside – but inside the carriage it is quiet and cool. I wonder who all these people are and where they are going? I don't normally ride this section of the line. I imagine that most of the Israeli Jews riding the line this far north live in settlements. I wonder what they think as they ride through Palestinian areas. Are they scared? Will anything happen? My ride is like most others. It is entirely peaceful, normal, boring, ordinary, comfortable. People just get on, find a seat, mind their own business and get off again. The conflict is not here.

This chapter recounts surprisingly comfortable encounters in a geopolitical context usually framed as anything but comfortable. Using my everyday experiences of riding Jerusalem's light rail, I offer an alternative writing of the geographies of the Israeli-Palestinian conflict, and challenge commonplace assumptions regarding the discomfort of living through a protracted low-intensity geopolitical conflict. In doing so, I seek to make space for a politics of (dis)comfort, arguing that comfort – and concomitant notions of indifference – often have profound geopolitical ramifications. The comfort of some, I argue, is often to the detriment and discomfort of others.

Between September 2012 and January 2014 I travelled to Jerusalem in order to carry out my doctoral research. I set out to explore the everyday geopolitical experiences of Israeli Jewish religious believers in a quintessential conflict city. I wanted to explore the ways and means by which the Israeli-Palestinian conflict was viewed and experienced through the lens of particular religious commitments, practices and theologies. This entailed in-depth ethnographic engagements with Jewish Israeli research informants in and throughout their quotidian lives. Previous ethnographic studies of everyday life in Israel position the conflict front and centre, emphasising the geographical omnipresence or potentiality of violent events and existential insecurity. These accounts portray Israeli Jewish lives as ones marked by discomfort and disorder; bodies that are forced to constantly negotiate disagreeable encounters or move through uncomfortable and precarious spaces. Fear of danger, pervasive suspiciousness and heightened alertness seem to saturate daily existence, informing even the most minor of decisions such as whether to stay in or go out for a coffee (Konopinski 2009; Ochs 2011).

In contrast, by affording due attention to the unspectacular moments that filled my research diary, this chapter pushes against the scholarly propensity to foreground particularly uncomfortable and extreme events and experiences, and the associated fear-based emotions or affects they engender (Pain and Smith 2008; Struckman and Sturm 2013). Most of the time I spent with research informants involved relatively comfortable experiences; drinking coffee in American style coffee-houses, shopping in malls, visiting 'tourist' sites, playing soccer, visiting friends, playing computer games, in addition to attending religious meetings and gatherings. Rarely were these shot through with discomfort, or emotions of fear, suspicion and trepidation. Even at times of heightened geopolitical tension, everyday life for my informants appeared to 'go on' as normal in relative comfort.

In order to explore this, I present a case study of one of Jerusalem's newest modes of public transport, the light rail. I draw on one unspectacular research diary entry (the extract that began this chapter) to show how the mundane use of the inter-city train reflects and shapes the geopolitical attunements of those who ride it. I initially use the story of the train to point to various contextual geopolitical debates that exist in the background of everyday life Jerusalem. These are well-trodden contestations over security, settlements, borders, territorial continuity, spatial segregation and ethnic social interaction. These are foregrounded in order to make a crucial observation: that whilst embroiled in wider geopolitical struggles, daily encounters between Palestinians and Israelis on the train did not appear to be completely defined by them. Instead, I explore the unexpected skein of daily encounters in the spaces of the train that seem – at first glance – to be relatively comfortable.

However, I move on to argue that these encounters were indicative of a much wider Jewish Israeli orientation of comfortable indifference – not simply towards Palestinian bodies – but to the Israeli-Palestinian conflict itself. Simply put, the conflict no longer seemed to play a significant or formative role in the ways Israeli Jews orient their everyday lives in space and time. To admit this seems counter-intuitive in light of the devastating violence that is often held as motif-like of the Israeli-Palestinian dispute. Here, the unequal distribution of violence and its diminishing impact on Jewish Israeli life, combined with the post-2005 unilateral disengagement from Gaza, and a reduction in Palestinian 'terror' attacks, has led to the Israeli-Palestinian conflict being pushed further into the background of Israeli public consciousness.[1] Comfortable indifference, it seems, is increasingly becoming a powerful geopolitical orientation in Israel.

Jerusalem's light rail

> I am riding the train back home from El Sal to the Damascus Gate after visiting a friend in Shuafat…

The light rail is the newest and fastest mode of urban transport in Jerusalem. At the time of research, the train had been fully operational for just over a year. The eight-mile-long track – roughly a 50-minute journey – cuts sleekly through the traffic of some of Jerusalem's busiest roads, servicing over 23 stations from Pisgat Ze'ev in the north to Mount Herzl in the west.

During construction, the line was beset by prolonged delays due to technical faults, budgetary and staffing issues and disruption caused by archaeological findings along the route (Sherwood 2014). As well as the controversy caused by construction delays, the line garnered significant political criticism. In official representations, the train was presented simply as an infrastructural necessity for alleviating Jerusalem's notorious traffic congestion. However, as Nolte and Yacobi (2015) argue, the train and its accompanying infrastructure can be viewed as spatial representations of a set of very particular socio-political relations. Despite the official functionalist representation of neutrality, modernism and efficiency (Busbridge 2014; Nolte and Yacobi 2015), the building of the line has been politically divisive and feeds into a number of geopolitical debates that feature heavily in critical geographical work about Israeli-Palestine.

Firstly, the route of the train line caused controversy for the simple reason that it connects areas of West (Israeli Jewish) Jerusalem to East (Palestinian Arab) Jerusalem. Accordingly, many Palestinians view the line as a type of 'infrastructural geopolitics' (see Weizman 2007) through which Israeli control and sovereignty will come to be extended and reified throughout the city (Nolte and Yacobi 2015). Jabareen (2010), Busbridge (2014) and Nolte and Yacobi (2015) all make the argument that the train is part of an on-going territorial 'master plan' that will extend, expand and normalise Jewish jurisdiction over the greater Jerusalem area through endurable infrastructural projects, the confiscation of Palestinian lands, and the exclusion of Palestinians from strategic urban planning decisions. Specifically, the line connects the centre of Jerusalem with surrounding Jewish settlements. The train line effectively smoothes space, giving the vague impression of a unified and uncontested cityscape. Hence, Busbridge (2014: 95) argues that the train 'entrenches the occupation insofar as it enacts the Israeli claim to "united Jerusalem"'.

Secondly, the train can be viewed as a curious form of mobile border/borderland as it simultaneously both contests and concretises existing geopolitical boundaries. As Elden (2013) notes, a significant portion of the rail line runs parallel to – or directly on top of – the Green Line.[2] For many Israeli Jews, the Green Line no longer holds any semblance of geopolitical meaning as a 'consistent differentiating marker' (Leshem 2015:36). The train helps exacerbate this growing obscurity/invisibility by running directly over its former route.[3] At the same time, the train also does the work of materialising and re-demarcating the Green Line and its urban logic of separation – albeit in a 'soft' way. As Pullan *et al.* (2007: 178 my emphasis) suggest, the train is not simply '*on* the border, it *is* the border'. Here, the train concretises Jerusalem's ethnic division between East and West more lucidly.

Lastly, the train is a form of public transport that exists uncomfortably within the Israeli state's wider (im)mobility regime. This regime seeks to restrict the movement of Palestinians through a myriad of interconnected measures. These consist of a combination of physical infrastructural obstacles – walls, fences, earthworks, roads and tunnels – and bureaucratic and biopolitical controls including a complex system of passes, permits and acts of closure (Pullan et al. 2007; Harker 2009). One consequence of this (im)mobility regime is the almost complete separation of Israeli and Palestinian modes of transport. In the everyday fabric of urban life in Jerusalem, this ethnic separation is manifest most obviously in the separate bus systems used by Israelis and Palestinians (Romann and Weingrod 1991). And yet the train appears to stand in contrast to this as it is available for use by both Israeli Jews and Palestinians (with the correct Jerusalem identification or permit).

In light of these wider geopolitical controversies, it stands that the train is a highly contested infrastructural element in a deeply conflicted and segregated city. It would, therefore, be unsurprising for the light rail to be interpreted primarily through the lens of a geopolitics of conflict, and concomitant notions of discomfort and antagonism. One might expect the train to be a site marked by particularly uncomfortable interactions; it is, after all, one of the only sites where Israelis and Palestinians come into proximate and sustained contact with each other in everyday life in Jerusalem. (Nolte and Yacobi 2015). Indeed, the owners of the train line also anticipated theses uncomfortable encounters. Planners were initially disinclined to direct the tracks through the Palestinians areas of Shuafat and Beit Hanina for fear of deterring Israeli Jewish passengers, and the chance of increased terror attacks (Sherwood 2014). During construction, the owners undertook a survey to ascertain Jewish Jerusalemites' nascent attitude towards the new train. The survey asked respondents to rank their concern on the following questions 'The light rail includes three stations in Shuafat. Does that present a problem for you?' and 'All passengers, Jewish and Arab, enter the train freely and without the driver's inspection. Is that a problem for you?' (Hasson 2010).

The everyday train

And yet, while cognizant of these broader volatile geopolitical contexts, the everyday encounters I observed on the train did not seem to be entirely indicative of them. My expectations for explicit manifestations of conflict were totally unfounded. In over 50 hours of riding the train, I observed no overtly conflictual encounters. More than that, I rarely recorded bodily gestures, or spoken articulations, of fear or discomfort in my observational research diary. Its atmosphere was extraordinary because of its normality and apparent comfort.

For this reason, I have found it helpful to think of Jerusalem's light rail as a type of Foucauldian 'heterotopia' (Foucault 1984). For Foucault, heterotopias are particularly elusive spaces; spaces that defy singular interpretations and resist closed normative understanding of the social interactions that happen within. They are spaces that allow for, and by their nature, enable a different set of social rules than those in surrounding social norms. Jerusalem's train acts as a type of heterotopia because of its apparent ability to disrupt the ethnic segregation and ethnicisation of everyday life through the sudden presence of 'complexity, contradiction and diversity' (Boedeltje 2012: 1). As previously intimated, in Jerusalem, the ethnic identity of places and people marks space; practically everything is designated as either 'Arab' or 'Jewish', and individuals frequent these spaces accordingly. Yet the train is seemingly for 'equal benefit of all the people in Jerusalem, Israelis and Palestinians altogether' (Nolte and Yacobi 2015: 29).

As the extract below suggests, the train facilitated the frequent, sudden and proximate presence of the 'Other'. *The Guardian's* Jerusalem correspondent, Harriet Sherwood (May 29th, 2014), describes it in the following terms:

> Men dressed in ultra-orthodox monochrome, under hats and coats even in the Middle Eastern summer, squeeze on board, averting their eyes from young women tourists in shorts and skimpy t-shirts. Religious Jewish mothers, hair bound in long winding scarves, with a brood of small children clutching at their ankle-length skirts, stand alongside Palestinian women in skinny jeans and elaborate hijabs framing carefully made-up faces and groomed eyebrows. Israeli soldiers in uniform, some armed with guns and all apparently armed with smart phones, lounge on seats opposite Palestinian labourers heading for jobs in Jewish areas of the city. Christian pilgrims en route to Via Dolorosa and the Church of the Holy Sepulchre, the site of Jesus's crucifixion, mingle with Muslims heading to the sacred Dome of the Rock and al-Aqsa mosque, and Jews intent on praying at the revered Western Wall…

The boundaries between disparate bodies are ruptured as individuals collide into one another when the train brakes suddenly or swings around a long bend. Such close bodily proximity between Israeli and Palestinian is rarely tolerated in other parts of public life, but the train facilitates an unintentional intimacy that is not found in many other everyday spaces in Jerusalem (Romann and Weingrod 1991; Sherwood 2014). One is reminded of Foucault's analogy of the cemetery as a form of heterotopia. Here, the meeting between the bodies of the living and the dead is confined to, and acceptable only in, that unique space.

Certainly, the materiality of the train facilitates these bodily encounters. Whilst there are seating areas, these are not prominent. Instead, the carriages are relatively open-plan, allowing for an abundance of standing room. Individuals find themselves sitting or standing next to people they might not normally choose to.[4]

> It's rush hour, and the already crammed train stops at the City Hall station. People fight to get out, whilst others fight to get in. It is impossible not to get carried down into the carriages in the sudden momentum of moving bodies. There is no choice but to stand and no choice who to stand next to.
>
> (Research Diary, May 9th, 2013)

But if these encounters do not appear to lead to antagonism or violence, can one express optimism at this seemingly peaceful 'contact zone' (Askins and Pain 2011: 803). Some argue that the train does appear to be an unexpected site of contact that may allow for individuals to encounter plurality and possibility. Busbridge (2014: 96), for example, states that the train 'has emerged as one of a few sites of tentative coexistence unthinkable only a handful of years ago'. She goes on to suggest that 'it is the possibilities of interaction and the reality of tentative coexistence that offers slivers of hope in an otherwise bleak landscape' (Busbridge 2014: 78).

However, I suggest that the train stands as a space of illusion to those who would herald it as site formative of harmonious geopolitical orientations. Spend enough time riding the line and one becomes aware that these comfortable encounters between Israeli Jews and Palestinians are marked by near-total indifference. The remainder of the chapter uses the train to explore everyday orientations of Israeli indifference in order to parcel out a politics of comfort. I initially examine individual Israeli indifference towards the presence of Palestinian bodies, before exploring a wider societal orientation of comfortable indifference towards the Israeli-Palestinian conflict.

Encountering indifference

> ...My ride is like most others. It is entirely peaceful, normal, boring, ordinary. People just get on, find a seat, mind their own business and get off again...

I suggest that the spaces of the train simply allowed for (non)encounters between Israeli Jews and Palestinian Arabs that were 'mooded' by an unstable negotiation of acknowledgement and indifference (Haldrup *et al.* 2008: 118, Makdisi 2010: 526, Sherwood 2014). In micro-acts of 'voluntary segregation', both Israelis and Palestinians appeared to pay little or no attention to the 'Other' in the carriage; they were 'so close to one another in actual physical space, [yet] …seem to be on different planets' (Romann and Weingrod 1991: 5). Ash Amin (2013: 4) describes the scene well:

> Strangers mingle or communicate with a degree of disinterest in each other, loyal to themselves, particular goals, and intimate others in and beyond that space…

These prosaic and unspectacular (non)encounters are rarely acknowledged in studies examining interactions between conflictual 'Others'. Instead, fear-filled, confrontational or uncomfortable moments of contact are favoured (Shirlow 2008). Yet these are the everyday exception rather than the rule in Jerusalem. My research diary is replete with mundane normality and (non) encounter, and to ignore such moments would be to dismiss the majority of everyday life.

Once I had observed indifference in the spaces of the train carriages, I began to observe it in many other sites in Jerusalem, in malls, in cafes and on construction sites. The train, it seemed, was simply a microcosm reflecting similar forms of social avoidance and disengagement occurring in spaces across this contested – yet fully functioning – city (see Romann and Weingrod 1991).

These (non)encounters fundamentally challenge certain geopolitical imaginaries of Jerusalem as a highly contested city divided by complete ethnic *hafrada* (Nolte and Yacobi 2015).[5] It was not unusual, for example, to hear my Jewish Israeli informants make the clichéd maxim that it was possible to live everyday life 'without ever coming into contact with a Palestinian'. Yet, as my research progressed, it became clear this was a rhetorical construction that masked the reality that Palestinians populate the everyday lives of Jewish Israelis. Far from being totally absent, my informants had simply learnt to ignore Palestinian bodies. Israelis, Romann and Weingrod (1991: 3) suggest, often know very little, if anything, about the Arab individuals who occupy their daily lives, allowing them to 'live together separately'.

In fact, present-day Jerusalem displays 'a tentative, albeit asymmetrical and uneven, mixing as Palestinians and Israelis encounter each other more often in the urban fabric of the city' (Busbridge 2014: 78). Palestinian bodies are more visible and more present in the spaces of West Jerusalem than ever before (Busbridge 2014). Aided by the train, Palestinians increasingly travel to the commercial spaces of West Jerusalem in order to work and enjoy leisure facilities.[6] Consequently, instances of contact occur every day in a myriad of interactional settings beyond the train carriages; individuals are drawn together and entangled in regular contact at the mall, in city parks, coffee shops, hospitals, offices, garages, at the 'ethnic borders' of residential neighbourhoods, on construction sites (Romann and Weingrod 1991).

Whilst I do not wish to downplay certain acts of divisive, systematic and thoroughgoing ethnic segregation that occur in Jerusalem, it does not take long to realise that geopolitical imaginaries of complete separation are only part of the picture. As Busbridge (2014: 81) similarly observed, 'instances of "clear-cut separation…are not only anomalous, but largely illusory"'.

Israeli Jewish indifference and the Palestinian other

It was, therefore, an important political act for my informants to claim never to have encountered Palestinians. It is significant that Israelis live within the comfortable bubble of their own socio-spatial worlds, constantly imagining that the 'Other' does not exist. But how is space rendered devoid of uncomfortable encounters with Palestinians? Or, in the words of Israeli anthropologist Jeff Halper (2008:39), 'how do we [Israeli Jews] render Palestinians invisible in a tiny territory where some 5.5million Jews live cheek by jowl with almost the same number of Palestinians?'

Halper (2008) goes on to suggest that Israeli indifference to the Palestinian Other is made possible by the presence of what he calls a 'cognitive membrane' that acts to screens out 'anything not having to do with them [Israeli Jews]' (Halper 2008: 66). Because of this filter, the Palestinians fade into 'mere background', rendered 'entirely irrelevant if not invisible' and 'dismissed and ignored' (Halper 2007: 36–44; Hanafi 2009; Vick 2010). Halper's observations touch on two important points pertaining to Israeli Jewish and Palestinian (non)encounters of indifference. Firstly, Halper's (2007: 44 my emphasis) notion of a '*cognitive* membrane' implies that indifference relies – in some small way – on cognitive work in the site of encounter. Individuals choose what to ignore, a decision embodied and enacted through the turning away of the body or the aversion of the eyes.

Here, comfortable indifference is understood as a conscious and spatial practice, one 'lived over and over in the glancing encounters of the street' (Tonkiss 2003: 300). However, I would argue that there is also a pre-cognitive or pre-reflective aspect to the everyday indifference that I observed on the train. Through small repetitive spatial practices – such as the ones described in the spaces of the train – insouciance comes to function at a more automated, habituated and subconscious level. It governs encounters in ways that individuals become less and less aware of, and, to a certain extent, have little control over. Through automaticity, one learns, as Tonkiss (2003: 301) observes, 'to look past a face'. *Contra* Romann and Weingrod (1991: 220), encounters with ethnic difference are not 'constantly in mind' in Jerusalem. This gives too much credence to the process of cognitive noticing. Over time, practised indifference solidifies into more complete detachment, denial and inattentiveness through embodied experience of daily life; positions that routinely bypass the cognitive or conscious (Cohen 2001).

Halper (2007) also implies that indifference can come to act as a collective societal geopolitical orientation (Bar-Tal 2001). Put otherwise, whilst indifference, denial, noticing/ignoring are 'neither fixed psychological "mechanism[s]" nor universal social process[e]s', they are never simply individuals acts (Cohen 2001: 3). Instead, as Zerubavel (2006: 20) argues, the individuals and objects to which we afford our attention are always grounded in much wider 'social traditions of paying attention'. Hence, what we come to dismiss or ignore is done as members of historically situated social communities; we are unmistakably socialised to pay attention to particular things in specific ways.

Indeed, the social aspect of what we perceive or ignore is made evident in the ways that this changes over time. As wider social attitudes shift, so does our collective focus (Zerubavel 2006). The outbreak of the second Intifada, for instance, made it impossible for Israelis to maintain the indifference that was more evident in the Oslo Accord years. Instead Israeli awareness of the Palestinian Other was considerably heightened; their presence became the source of fear, anxiety and discomfort. Accordingly, the Israeli citizenry were marked by a hyper-vigilance that meant that the Arab subject was under constant gaze. Today, it is more likely that Israeli Jews 'walk by, unperturbed by the silent gaze of their Arab neighbors' (Pinto 2013: 78).

Jewish Israeli indifference and the Israeli-Palestinian conflict

'...The conflict is not here...'

Thus far, I have argued that small, habituated practices of everyday insouciance are both reflective and formative of a wider societal disposition of comfortable indifference towards the Other. However, this indifference did not simply mark my informant's orientation towards Palestinians. Crucially, at the time of research, I would argue that comfortable indifference had become a much wider geopolitical orientation that conditioned Israeli Jewish everyday (dis)engagements – not simply with the Other – but also with the topic of the Israeli-Palestinian conflict.

We are used to thinking of Israeli society as being marked by a siege mentality; rife with fear, heighten anticipation and security obsessions.[7] But, a shrugging of the shoulders is part of a political *raison d'etre* that dominates Israel today. Crucially, none of my informants ever talked at length about feeling afraid of the conflict when going about their everyday lives. Instead, after over 50 years of occupation in the West Bank and Gaza, two Palestinian intifadas, several periods of intensified violence, a unilateral disengagement from Gaza and numerous failed political processes, the Israeli public appears to have become entirely disinterested in the on-going conflict. Of course, casting all Israelis as indifferent in this way is a simplistic and homogenising caricature. I certainly do not apply this definition in a rigid manner, or argue that it applies to every Israeli.[8] But political commentators in Israel are beginning to recognise that Israeli interest in the conflict is waning.

Certainly, academic focus on Israeli indifference also appears to be increasing (see Natanel 2016). This is most evident, for example, in the on-going and prevalent social-psychological literature addressing the effects of the conflict on Israeli society. Once this literature would be rife with accounts of societal fear. However, psychologists Greenbaum and Elizur (2012), and Halperin *et al.* (2010: 67) all contend that Israelis have disengaged from issues pertaining to the conflict through a myriad of psychological defense behaviours that combine elements of 'justification, rationalization, and dissonance reduction' (Greenbaum and Elizur 2012: 396). Mechanisms such as repression, denial, projection, rationalisation and avoidance 'allow [Israelis] to avoid facing the contradictions between their group's behaviors and the moral values that are acceptable in modern societies'. Halperin *et al.* (2010: 62) observed that comfortable indifference has come to affect 'a large majority of [Israeli] society members…and provides an orientation for the group's behavior in the context of occupation'. Most recently, anthropologist Helman (2015) diagnosed a 'culture of collective denial' within Israel vis-à-vis the conflict. Whilst I would suggest that the term 'indifference' is more accurate than 'denial' (due to reasons of conscious/unconscious (un)intentionally), the baseline sentiments of a collective societal orientation marked by 'unnoticing' remain the same.

An everyday geopolitics of comfortable indifference

Whilst these studies point towards the socio-psychological state of Israeli indifference, they do not proffer many suggestions as to *why* or *how* this condition occurs. What are the underlying causes for this collective orientation of indifference? Clearly it must be the result of a number of processual phenomena, the range of which is far beyond the scope of this chapter. However, I examine some of the primary causes, and go on to argue that these are constitutive of a geopolitics of comfortable indifference.

Of course, for many, the situated immediacies of everyday life simply 'get in the way', causing one to forget about both the presence of the 'Other' and the wider conflict. After being shocked at witnessing an Israeli military court for the first time, Israeli journalist Alon Idan (2011: n.p) confessed:

> Here's an example of the way you forget: At a certain point on the journey home [from the military court], a car behind me flashed its bright lights, urging me to switch lanes… For the next few seconds, I'll think about the barbarism of the driver who used the bright lights. That's where my thoughts will be. And then someone will call, and we'll talk about sports until I get home and park. When I enter my house, I'll give my child a bath and then I'll lie down to go to sleep. Later, I'll fall asleep. Tomorrow is another day.

But apart from the immediacies of everyday life, are there other geopolitical causes that can be found for this growing orientation of comfortable indifference amongst the Israeli public? To my mind, it is clear that spatial and discursive strategies of separation and security are causative of Israeli indifference (Yiftachel 2005). As Azoulay and Ophir (2013: 12) attest, 'the ongoing [Israeli] control of the Occupied Territories is conceived of [in everyday political discourse] as incidental and, especially, external to the Israeli regime'. To enable the normality of everyday life, 'the Occupied Territories are bracketed off, forgotten, and denied' (Azoulay and Ophir 2013: 18). Discursively, for instance, the conflict is veiled and distanced behind certain lexical rhetoric; it is simply a 'situation' that occurs – for the most part – in the abstract geopolitical legalese of the 'disputed territories'.

More concretely, actual hostilities are often undetectably distant,[9] taking place out of sight and hidden behind the concrete walls of the separation/security barrier/fence/wall. Here, Israeli indifference grows as the conflict is rendered gradually invisible and distanciated by security apparatus (Neve Gordon 2008). As Azoulay and Ophier (2013: 17) claim, 'Most Israeli citizens...usually enjoy the privilege of suspending the Occupation's violent presence, distancing it from sight and heart and forgetting it exists'. Similarly, one of my informants stated, the conflict 'may as well be happening thousands of miles away'. *Contra* Ochs (2011) and Konopinski (2009), fear of danger no longer saturates Israelis' everyday lives. Instead, Israel's vast and complex security regime has allowed Israeli citizens to become increasingly comfortable – perhaps even numb (Pinto 2013).

This begs the questions: what wider impact does indifference have on the conflict? What geopolitical practices does this comfortable indifference facilitate? Does anyone benefit from indifference? Can it be managed or manipulated? In asking such questions, we have come a long way from those encounters on the train. Comfortable indifference is no longer presented simply as an everyday reaction. Rather, I want to suggest that it is a constituent part of the political apparatus. More than simply a reaction, indifference can come to be enrolled as an operational element within certain systems of power.

Confident in their political and military power, Israeli indifference effectively secures the 'current pleasant status quo (for Israeli Jews)' (Pinto 2013: 190). Indeed, Halperin *et al.* (2010), Greenbaum and Elizur (2012) and Bar-Tal *et al.* (2010) all imply that current Israeli indifference results in shoring up and perpetuating the conflict. I would suggest that this occurs in two overlapping ways. Firstly, indifference can lead to and legitimise violent practices through a lack of social critique, and secondly, indifference closes down the possibilities for peace. These will be taken in turn.

Firstly then, public indifference can so often be the basis from which State-sponsored acts of reactive or pre-emptive violence are mobilised, performed and justified. Wider indifference acts as a safety net of denial that translates into public legitimacy and, in turn, relative impunity (Cohen 2001). Crucially, indifference erodes one's impulse to know about the violence that is done in one's name. Woodward (2013: 102) states the problem succinctly, arguing that 'apathy and indifference for the Palestinian cause could easily transition into…unconcern when military options are employed'. Accordingly, psychologists Greenbaum and Elizur (2012: 396–398) suggest that there is real 'cause for concern' about the lack of Israeli public engagement because this signals 'apathy born of dissociation and justification…toward the suffering of others'. They go on to observe that 'when all is quiet, there is little regard to what is happening to people in the OPT, and [Israeli] life continues as normal' (Greenbaum and Elizur 2012: 395). Most Israelis, Halper (2008: 66) claims, 'do not know a thing about either the realities on the ground in the Occupied Territories or what Palestinians think and want (and don't really care)'. With violence rendered firmly, 'out of mind', it is easy to see how Israelis quickly 'lose interested in the lives of the Palestinians' (Gordon 2008: 212). Whilst I agree with Benvenisti's (2009: n.p) sentiments that 'the situation in the occupied territories interests Israelis only when something violent takes place there', he does not qualify just how violent an event it now takes to capture Israeli interest.[10] After all, as UNOCHA[11] weekly reports illustrate, daily instances of violence persist unabated and largely unnoticed in the Occupied Territories.

Societal indifference has also been linked to escalations and cycles of violence and fear. For example, Greenbaum and Elizur (2012: 396) observe that:

> the greater the repression [of the conflict] by the occupier, the greater the violence by the occupied peoples, leading to…intrasociety violence, and increased use of psychological defense mechanisms. We suggest that these processes are the basis for the cycles of violence that have been described by observers of the Israeli-Palestinian conflict.

Similarly, the Israeli political scholar Neve Gordon (2009: 242) suggests that it is 'lack of interest or indifference to the life of most of the colonized population' that 'helps to explain the recent surge in lethal violence in the Occupied Territories'. It is becoming clear that in recent years Palestinians have been caught in a cyclical paradox as a result of periods of Israeli indifference. During periods when Palestinian armed groups were committing acts of violence, the Israeli government refused to engage with Palestinians on account of the 'terrorists'. However, with diminishing levels of Palestinian 'terrorism', the Palestinian 'issue' all but disappears from the Israeli agenda. Terrorism, it seems, made an otherwise invisible population visible. As Meron Benvenisti (2009: n.p) contends, 'ignoring the situation is convenient for everyone [right and left wing Israelis], and therefore *all are partner to the concept that the Arabs are interesting only when they are violent*'. This was emphasised by a popular Israeli slogan that was prevalent during the Second Intifada 'no Arabs, no terrorism'. However, this works both ways; for many Israelis, 'no terrorism' equates to 'no Arabs'. As

Uri Misgav (2013: n.p) states, 'when they shoot, you can't talk about peace. When they don't shoot, why should you talk about peace?'

Consequently, my Palestinian friends felt that they were left with few viable options for changing the status quo. At best Israeli indifference will force the increasingly disempowered Palestinian Authority into making unilateral political moves. This was most evident during my research with the push for – and acceptance of – Palestine as an observer state at the United Nations in 2013. At worst, Palestinians could point to Israeli indifference as a justification for revisiting more violent means. Then, Israeli indifference will be shattered by another Intifada or associated violent hostilities (indeed, see Coda).

Manipulating/manufacturing comfortable indifference

In critical geopolitics, it is often claimed that fear or moral panic are furthered – through a variety of political devices and mechanisms – to advance and justify certain geopolitical policies/practices or bolster incumbent administrations (Pain and Smith 2008; Kirby 2013).[12] However, is there a case for arguing that comfortable indifference can also be manufactured, manipulated or mobilised towards certain geopolitical ends? To my mind, there are some for whom an Israeli public consciousness steeped in indifference is desirable, or – at the very least –advantageous.

Certainly, for the current ruling parties, Israeli public indifference is valuable because it acts as a fail-safe that militates against the need to make significant or immediate geopolitical concessions.[13] Cohen (2001) points to the ways in which an indifferent public is also a public who lack the preparedness, capacity or willingness to make significant political changes. As history has shown, the political and electoral cost of reaching a peace agreement in Israel can be high. Halper (2008: 88) illustrates this in his description of the sudden shattering of indifference that accompanied the signing of the Oslo Peace Accords in 1993:

> Without being prepared in any way, Israelis turned on their TVs one day in September...and saw their Prime Minister shaking hand with the person whom they had been told for a generation was the arch-terrorist, the ultimate foe of Israel...Suddenly their whole world was turned upside down...within five months the Israeli public voted in Rabin's antithesis, the very personification of the security framing, Benjamin Netanyahu.

Similarly, Woodward (2013: 96) recently argued:

> The price [of peace] to be paid, both now and later, is greater than the benefit to be realized now. There is regrettably a short-term mindset pervasive in Israeli culture and government. The goal has been, and continues to be, conflict management, maintaining the status quo in

> perpetuity, and not conflict resolution. All policy is geared toward preserving and sustaining a functional normality. Israel thus plays an ad hoc game of meeting each challenge as it arises, keeping the lid on the boiling pot knowing that the heat will not be completely turned off.

All this to say that, at present, the status quo appears to be viewed as being more advantageous. Under Netanyahu's direction, there has been a consolidation of a gradual shift in Jewish public consciousness from 'conflict resolution to conflict management' (Yiftachel 2005:127, Blumenthal 2013). Netanyahu has advanced this quietist policy of conflict management predicated on the rhetoric strategy of 'neither war nor peace' or 'peace without peace' (Blumenthal 2013: 401). Here, the conflict is 'carefully managed – but never ended – to guarantee tranquility for Israeli Jews…As long as the one-way peace holds, Israelis [will] support the status quo, and by extension, Netanyahu' (Blumenthal 2013: 358).

Crucially, Israeli indifference can be manufactured by disseminating the seemingly dialectical ideas that, on the one hand there is no political solution to the conflict, and on the other, that it is only temporary and forever on the verge of being resolved (Azoulay and Ophir 2013). Halper (2008: 65) points to Ehud Barak's hugely influential contention that Israel had 'no partner for peace', a phrase coined by the Israeli minister after the failure of the Camp David Peace Talks. From that point on, the Israeli public – deprived of a permanent political alternative/solution – disconnected 'itself from the political process' and developed a 'bunker mentality' (Marzano 2013: 96). All that was left to do was to 'hunker down' in an introspective geopolitical bubble, to simply 'get on with our lives' (Halper 2008: 65).

However, at the same time, the Israeli public is fed the illusion of the temporariness of the conflict, and the deception of an impending possible peace (Azoulay and Ophir 2013). This amounts to what Yiftachel (2005: 128) terms a 'politics of suspension'. Joseph Massad (2014) lucidly points out that when the Madrid peace talks began in 1991 – the occupation of the Palestinian Territories had existed for 24 years. To date, therefore, there have been 28 years of peace negotiations to end a 24-year-old occupation. As a result, Israelis are orientated away from taking responsibility for peace as it is purportedly just around the corner. In reality, the Israeli state continues to stall or freeze peace processes through various means. As Woodward (2013) states,

> At times, the parties have come to the brink of peace and then pulled away, with each occasion serving only to reinforce that feeling of futility, leading to pervasive resignation, even apathy, in much of the general public…this may be more prevalent in the Israeli public…

The sum of these two dialectical narratives is the psychological condition of 'permanent impermanence' (Ginty *et al.* 2007: 3); a condition that numbs the

Israeli public to a protracted political standoff by keeping alive the ephemeral illusion 'that the occupation is about to end, and at the same time to convince the majority that this cannot possibly happen overnight' (Reinhart 2002: 226). As Azoulay and Ophir (2013: 14) claim, 'the false temporariness of "the Occupation" generates perceptual blindness that is at one and the same time caused by the ruling apparatus in the Territories and one of its active mechanisms'.

In this way, comfortable indifference is created and maintained. And, under the 'cover of this lack of interest, the Israeli rule over the West Bank is continuing to deepen' (Benvenisti 2009: n.p). Indeed, increasingly there are those who argue convincingly that an indifferent Israeli public allows for the extension of territorial expansionist projects in the West Bank (Azoulay and Ophir 2013). For example, Oren Yiftachel (2005: 128) argues that indifference allows for a 'game of deception' where 'all actors turn a blind eye and continue to support the illusion of impending peace'. Meanwhile, a long-term, strategic and domineering 'political geographic order best described as "creeping apartheid" is allowed to occur; an order where 'Jews continue to settle in the West Bank, the illegal wall is still being constructed, and the treatment of some groups among Israel's Palestinian citizens increasingly resembles the fate of their brethren in the Occupied Territories' (Yiftachel 2005: 128).

Hence, to conclude, as Azoulay and Ophir (2013: 17) claim:

> Israeli citizens of Jewish descent take part in and are ruled by the regime of which 'the Occupation' is one element; they contribute to its reproduction, not only as soldiers, settlers, or government officials, but also as its governed subjects, who tacitly accept its rules and perpetuate its legitimacy, mostly by ignoring how it rules others, non-Jews and noncitizens, letting it be inscribed and reinscribed in the movement of their bodies, the wording of their language, and the limited horizons of the political imaginations.

Conclusion

In this chapter I have suggested that comfort and indifference are often entangled with significant geopolitical results. It became increasingly clear that during the period of my research, the Israeli-Palestinian conflict rarely impacted the everyday lives of many Israeli Jews in Jerusalem. I have suggested that this extended period of calm had allowed a certain geopolitical orientation to predominate amongst Israeli Jewish individuals and society – one of comfortable indifference.

Far from being insignificant, I demonstrated that a societal orientation of comfortable indifference has both geopolitical causes and effects. Distancing the conflict from the consciousness of the Israeli populace relied upon certain geopolitical imaginaries, infrastructures and praxis. Crucially, like moral panic or widespread fear, I suggested that orientations of public indifference could be used as a platform for statist political manoeuvrings. I argued that Israeli Jewish indifference has allowed the state to enact various geopolitical agendas; predominantly the continuation of the status quo of occupation and the creeping colonisation of the West Bank. However, I have also suggested that Israeli indifference would eventually encourage more violent responses from Palestinians. History shows that periods of relative calm – such as the one I experienced during my fieldwork – rarely last; Israeli indifference will always come to be shattered.

This also signals a broader shift in the interrogation of the geopolitics of comfort and indifference not as a specific condition of the Israeli-Palestinian conflict, but one that is perhaps a configuration of contemporary low-intensity conflict more broadly. What role, for example, did Western comfort and indifference play in the composition and continuation of the Bosnian, Iraq or Afghanistan conflicts. More recently, how did/does our indifference influence our (dis)engagement with significant events such as the Ebola outbreak or the Syrian refugee crisis? Whilst critical geopolitical scholars are quick to point out that our negative imaginaries of people and places will be formative of geopolitical praxis, is it not also true that a lack of imagination will be significant?

Coda

Throughout 2014 – the year after my research – Jewish Israeli comfort and indifference was momentarily shattered. In the summer, a prolonged and deadly war in Gaza led to an autumn that saw a spate of terrorist attacks across Israeli cities. At the time of writing, nine fatal attacks had occurred, along with a small number of non-fatal attempts. Increasingly, it became clear that these attacks were being carried out by lone East Jerusalem Palestinians with little organisational affiliation. Soon, protests, riots and clashes between Palestinians and Israeli forces became frequent. The 'security calm' that had come to characterise everyday life after the ending of the Second Intifada had been 'irreparably disrupted' (Benn 2014: n.p). Jerusalem, according to one of my informants, 'felt' tenser than it ever had done.

In the summer of 2014, in response to the kidnapping and brutal killing of a Palestinian boy from East Jerusalem, protestors targeted the infrastructure of the light rail, destroying ticket machines and signalling mechanisms, buckling tracks and throwing rocks at passing trains. The train service was disrupted for weeks. In the autumn, the lack of security – and the open-planned architecture of the station-stops – allowed for two separate terrorist attacks to occur within a month. In the first, a vehicle was driven along the light rail platform, killing and wounding several waiting passengers. In the second, a vehicle was used to ram the train. Protestors targeted the train because it concretised what they saw as creeping Israeli colonial territorialisation.

Notes

1 Here I am referring to hostilities that would directly affect Israeli society. Violence and insecurity mark everyday Palestinian lives in much more direct ways – see, for example, Hammami's (2006) work.

2 Or is built on the 'no-man's land' frontier of the Israeli–Jordanian border before 1967.

3 During my time in Jerusalem, an Israeli left-wing NGO painted the Green Line on its route through the city. They began metres from the train line on Jaffa Street – see https://www.youtube.com/watch?feature=player_embedded&v=2KQkjvyMubA

4 Other sites include hospitals (see Romann and Weingrod 1991), which Foucault (1984) also uses as an example of a heterotopia.

5 *Hafrada* is the Hebrew word for separation.

6 The reverse is not occurring (Sherwood 2014). It is still rare to see Israeli Jews (who are not 'settlers') in East Jerusalem.

7 I certainly do not wish to downplay these formative collective societal emotions. As Bar-Tal (2001) demonstrates, fear is a deep and historically powerful force in Israel; an emotion that can instantaneously override and flood one's consciousness due to a number of conscious or pre-cognitive cues.

8 To do so would brush over a profound array of political engagements and socio-psychological positions within Israeli society due to the protracted conflict (Halperin *et al.* 2010).

9 This distance, however, can be manipulated; its importance can be strategically overstated – or underplayed - at specific times. For example, distance becomes vitally important during peace talks or negotiations over future borders. In 2011, in response to Obama's increasing insistence that a just peace would be based on the borders of 1967, Netanyahu argued that these were not 'defensible borders'. Much was made of the fact that - suddenly - the 'situation' was occurring only seven miles away from Tel Aviv. The usually distant conflict was swiftly made near in order to rebut Obama's suggestions.

10 My Israeli informants seemed only to be interested in happenings in the West Bank if the violence directly affected Israeli Jewish settlers or army personnel.

11 United Nations Office for the Coordination of Humanitarian Affairs.

12 In the context of Israel, it is often suggested that a distinct politics of fear is at play where Israelis are perpetually reminded of a continual, existential threat in order to foster a semblance of political compliance (see Bar-Tal 2001).

13 Public indifference is just one of many such fail-safes in the Israeli political system that will make territorial concession almost entirely impossible (Spruyt 2014)

Bibliography

Amin, A. (2013) 'Land of strangers', *Identities*, 20(1), 1–8.

Askins, K. and Pain, P. (2011) 'Contact zones: Participation, materiality, and the messiness of interaction', *Environment and Planning D: Society and Space*, 29(5), 803–821.

Azoulay, A. and Ophir, A. (2013) *The one-state condition: Occupation and democracy in Israel/Palestine.* Stanford: Stanford University Press.

Bar-Tal, D. (2001) 'Why does fear override hope in societies engulfed by intractable conflict, as it does in the Israeli society?', *Political Psychology*, 22(3), 601–627.

Bar-Tal, D., Halperin, E. and Oren, N. (2010) 'Socio-psychological barriers to peace making: The case of the Israeli Jewish society', *Social Issues and Policy Review*, 4(1), 63–109.

Benn, A. (2014). *Israel's next election is a referendum on Netanyahu.* [online] Available at: <http://www.haaretz.com/news/national/.premium-1.629709> (Accessed 03/12/2014).

Benvenisti, M. (2009) *Interesting only when they are violent.* [online] Available at: <http://www.haaretz.com/print-edition/opinion/interesting-only-when-they-are-violent-1.270961> (Accessed 26/02/2009).

Blumenthal, M. (2013) *Goliath: Life and loathing in greater Israel.* New York: Nation Books.

Boedeltje, F. (2012) 'The other spaces of Europe: Seeing European geopolitics through the disturbing eye of Foucault's heterotopias,' *Geopolitics*, 17(1), 1–24.

Busbridge, R. (2014) 'Frontier Jerusalem: Blurred separation and uneasy coexistence in a divided city', *Thesis Eleven*, 121(1), 76–100.

Cohen, S. (2001) *State of denial: Knowing of atrocities and suffering.* Cambridge: Polity Press.

Elden, S. (2013) 'Secure the volume: Vertical geopolitics and the depth of power', *Political Geography*, 34, 35–51.

Foucault, M. (1984) *The Foucault reader.* (ed) P. Rabinow. New York: Pantheon Books.

Ginty, R., Muldoon, O. T. and Ferguson, N. (2007) 'No war, no peace: Northern Ireland after the agreement,' *Political Psychology*, 28(1), 1–11.

Gordon, N. (2008) *Israel's occupation.* Berkeley: University of California Press.

Gordon, N. (2009) 'From colonization to separation: Exploring the structure of Israel's occupation', in A. Ophir, M. Givoni and S. Hanafi (eds.), *The power of inclusive exclusion.* New York: Zone Books, pp. 236–268.

Greenbaum, C. and Elizur, J. (2012) 'The psychological and moral consequences for Israeli society of the occupation of Palestinian land', in D. Bar-Tal and I. Schnell (eds.), *The impacts of lasting occupation: Lessons from Israeli society.* Oxford: Oxford University Press, pp. 381–408.

Haldrup, M., Koefoed, L. and Simonsen, K (2008) 'Practising fear: Encountering O/other bodies', in R Pain, and S. Smith (eds.), *Fear: Critical geopolitics and everyday life.* Aldershot: Ashgate, pp. 117–128.

Halper, J. (2007) 'From protest to resistance: The making of a critical Israeli', *Journal of Palestine Studies*, 36(3), 36–44.

Halper, J. (2008) *An Israeli in Palestine.* London: Pluto Press.

Halperin, E. Bar-Tal, D., Sharvit, K., Rosler, N. and Raviv, A. (2010) 'Sociopsychological implications for an occupying society: The case of Israel', *Journal of Peace Research*, 47(1), 59–70.

Hammami, R. (2006) 'Sustaining life in a geography of adversity,' *Prince Claus Fund Journal*, 13, 68–83.

Hanafi, S. (2009) 'Spacio-cide: Colonial politics, invisibility and rezoning in Palestinian territory', *Contemporary Arab Affairs*, 2(1), 106–121.

Harker, C. (2009) 'Student Im/mobility in Birzeit, Palestine', *Mobilities*, 4(1), 11–35.

Hasson, H. (2010) *J'lem train company asks passengers: 'Do You Mind Traveling With Arabs'?* [online] Available at: <http://www.haaretz.com/j-lem-train-company-asks-passengers-do-you-mind-traveling-with-arabs-1.309101> (Accessed 20/08/2010).

Helman, S. (2015) 'Challenging the Israeli occupation through testimony and confession: the case of anti-denial SMOs Machsom watch and breaking the silence', *International Journal of Politics, Culture, and Society*, 28(1), 377–339.

Idan, A. (2011) *Another country*. [online] Available at: <http://www.haaretz.com/weekend/magazine/another-country-1.356147> (Accessed 15/04/2011).

Jabareen, Y. (2010) 'The politics of state planning in achieving geopolitical ends: The case of the recent master plan for Jerusalem', *International Development Planning Review*, 32(1), 27–43.

Kirby, P. (2013) *A critical geography of the Homeland Security Advisory System (2002–2011)*. Unpublished PhD. Department of Geography, Royal Holloway University of London.

Konopinski, N. (2009) *Ordinary security: An ethnography of security practices and perspectives in Tel Aviv*. Unpublished PhD. School of Social Anthropology: The University of Edinburgh 2009).

Leshem, N. (2015) '"Over our dead bodies" Placing necropolitical activism', *Political Geography*, 45, 33–44.

Makdisi, S. (2010) 'The architecture of erasure,' *Critical Inquiry*, 36(3), 519–559.

Marzano, A. (2013) 'The loneliness of Israel. The Jewish state's status in International Relations', *The International Spectator: Italian Journal of International Affairs*, 48(2), 96–113.

Massad, J. (2014) *Joseph Massad on how 'Peace is War'*. [online] Available at: <http://mondoweiss.net/2013/11/joseph-massad-peace.html> (Accessed 23/08/2014).

Misgav, U. (2013) *Editing (out) the occupation. New America live events*. [online] Available at <http://www.ustream.tv/recorded/33533525>

Natanel, K. (2016) *Sustaining conflict: Apathy and domination in Israel-Palestine*. Oakland: University of California Press.

Nolte, A. and Yacobi, H. (2015) 'Politics, infrastructure and representation: The case of Jerusalem's Light Rail', *Cities*, 43, 28–36.

Ochs, J. (2011) *Security and suspicion: An ethnography of everyday life in Israel*. Philadelphia: University of Pennsylvania Press.

Pain, R. and Smith, S. (eds.) (2008) *Fear: Critical geopolitics and everyday life*. Aldershot: Ashgate.

Pinto, D. (2013) *Israel has moved*. Harvard: Harvard University Press.

Pullan, W., Misselwitz, P., Nasrallah, R. and Yacobi, H. (2007) 'Jerusalem's Road 1: An inner city frontier', *City*, 11(2), 176–198.

Reinhart, T. (2002) *Palestine/Israel: How to end the war of 1948*. New York: Seven Stories Press.

Romann, M. and Weingrod, A. (1991) *Living together separately: Arabs and Jews in contemporary Jerusalem*. Princeton, NJ: Princeton University Press.

Sherwood, H. (2014) *Jerusalem's light railway: Commuting with a rifle through the conflicted city*. [online] Available at: <http://www.theguardian.com/cities/2014/may/29/jerusalems-light-railway-commuting-with-a-rifle-through-the-conflicted-city> [Accessed 29/05/2014]

Shirlow, P. (2008) "Ethno-sectarianism and the construction of fear in Belfast, Northern Ireland," in R. Pain and S. Smith (eds), *Fear: Critical geopolitics and everyday life*. Aldershot: Ashgate, pp. 193–212.

Spruyt, H. (2014) 'Territorial concession, domestic politics, and the Israeli-Palestinian conflict,' in M. Fendius, O. Haklai and H. Spruyt (eds), *Democracy and conflict resolution: The dilemmas of Israel's peacemaking*. Syracuse: Syracuse University Press, pp. 29–66.

Struckman, L. and Sturm, T. (2013) 'Introduction: Interrogating conventional geopolitics of the Middle East', *The Arab World Geographer*, 16(1), pp. 3–9.

Tonkiss, F. (2003) 'The ethics of indifference: Community and solitude in the city,' *International Journal of Cultural Studies*, 6(3), 297–311.

Vick, K. (2010) *Palestinians, contained.* [online] Available at: <http://content.time.com/time/magazine/article/0,9171,2036142,00.html> (Accessed 20/12/2010).

Weizman, E. (2007) *Hollow land: Israel's architecture of occupation.* London: Verso.

Woodward, S. (2013) 'Why can't we all just get along? The Israeli-Palestinian conflict', *Mediterranean Quarterly*, 24(4), 92–108.

Yiftachel, O. (2005) 'Neither two states nor one: The disengagement and "Creeping Apartheid" in Israel/Palestine', *The Arab World Geographer*, 8(3), 125–129.

Yiftachel, O. (2013) 'Liberal colonialism? Israel's 2013 elections and the "Ethnocratic Bubble"', *Journal of Palestine Studies*, 42(3), 48–67.

Zerubavel, E. (2006) *The elephant in the room: Silence and denial in everyday life.* Oxford: Oxford University Press.

7 Homely comforts abroad

Navigating the comfort zone(s) within international student mobility

Laura Prazeres

Introduction

'Home' is a term that evokes a complex register of feelings and meanings. People ascribe a variety of feelings to the notion of 'home'. Blunt and Dowling (2006) have described home as 'both a place/physical location and a set of feelings' (p. 22). People *feel* at home as much as they move *to/from*, *with* and *between* home(s). 'Home' is a complex and multilayered notion that is concurrently social, affective and situated in multiple temporal spaces. Recent conceptual definitions of 'home' have veered from fixed and stable to fluid and mobile (Mallett, 2004; Blunt and Dowling, 2006). Scholars have argued that 'home' *is* mobile (Germann Molz, 2008; Ralph and Staeheli, 2011). As an affective space that can include material belongings and social relations, 'home' can travel with you.

In this chapter I approach the notion of 'home' from both a spatial and affective frame. In particular, I explore homely places and feelings within the context of international student mobility. Mobility can provoke a re-evaluation of the meaning of home (Nowicka, 2007). People on the move can gain a different perspective and understanding of 'home' through distance and time (Terkenli, 1995; Wiles, 2008). Ralph and Staeheli (2011) have argued that different types of migrants (e.g. tourist vs. refugee) will have differing experiences of home. Much of the literature on home and mobility has focused on tourist, refugees, labour migrants and global nomads, with little work on international students. I address the absence of work on international student mobility in geographies of home by examining the experience of international exchange students working and/or studying in the Global South. The contrast between the different living conditions in the country of origin and a new country of residence can arouse feelings and ideas of 'home' to the forefront. This is particularly the case for my Canadian participants for whom sojourning in a developing country often involves travel – both physical and imaginary – between contrasting socio-economic realities. Time spent away from home-places can spark a resurgence of emotions linked to those mundane objects, relations and spaces that pervade our everyday life, prompting us to re-enact home-making practices in new places. Home is thus created and re-created through mobility (Ahmed et al., 2003).

I examine in this chapter how international students recreate homely feelings during educational sojourns abroad. I focus on the theme of 'comfort' as a predominant narrative of students' international exchange experiences in the Global South. Central to these narratives are students' motivations to leave their 'comfort zone'. Participants are eager to remove themselves from their comfort zone in order to achieve self-growth and personal change. The main reason for studying or interning abroad is to 'escape' a physical space and psychological level of comfort and familiarity – the comfort zone – yet, while away from the comforts and familiarity of home students attempt to reproduce such comforting feelings of homeliness. I argue that during their sojourns abroad students periodically seek out 'zones of comfort' to alleviate stress and feelings of homesickness. I demonstrate through participants' narratives how these zones of comfort are spatial (places), corporal (familiar foods) and social (family and friends). I thus focus on the role of familiar and comforting places, foods and social relations in reconstituting international students' feelings of 'home'.

The comforts of home, the zones of comfort

The 'comforts of home' is an adage that resonates throughout my study. 'Home' is conventionally regarded as a secure, comforting and familiar space. Traditional ideas and meanings of home connote a safe, intimate and private domestic space offering a refuge from the uncertainties, stresses and risks perceived within the public sphere. The human geographer Yi-Fu Tuan (1975) has described home as a haven providing a safe space of care and security. Mallett (2004) has noted that such ideas of domesticity and comfort are rooted and situated in historical and cultural Western accounts of home and remain prevalent in current times. Such affective, place-based constructions of 'home' suggest conditions of fixity and immobility, yet meanings of home as comforting and secure emerge through mobility (Case, 1996).

Mobility can incite musings on the meanings of 'home'. As people move through space and time, feelings of comfort and familiarity can surface to the forefront. It is thus unsurprising that the value of 'familiar places is often brought home to us when we are removed from them' (Skey, 2011, p. 235). In her study of mobile families, Allen (2008) has found that ideas of 'home' were linked primarily to feelings of familiarity, security and comfort. Far from being relinquished in the place of origin, these feelings are transported and recast *en route* or in a new place. Indeed, the affective underpinnings of home are not abandoned with mobility. Germann Molz (2008) has noted that travellers carry with them 'the transportable sentiments of comfort, security, familiarity, and control' (p. 327). Wise (2000) considers home as a continuous process of 'creation of a space of comfort' (p. 300). Home is always a work in progress. In examining young people's experience of leaving the parental home for higher education, Kenyon (1999) notes the importance of comfort for undergraduate British students' sense of home.

I consider in this chapter notions of comfort and familiarity as material, social, corporeal and spatial.

Although this chapter focuses on the relationship between home and feelings of comfort and familiarity, it does not deny the presence of discomfort and estrangement (Ahmed, 1999) or even violence and oppression (McDowell, 1999). Home need not be either comfortable or uncomfortable, nor familiar or unfamiliar. Indeed, such affective binaries can coexist (Mallett, 2004). I want to acknowledge valuable critiques and critical geographies of home (Constable, 1999; Fortier, 2003; Blunt and Dowling, 2006; Brickell, 2012) but suggest that for middle-class students who voluntarily travel abroad, comfort and familiarity are still central features of 'home'. Notably, students seeking a challenge choose to embark on a sojourn abroad to escape the comforts and familiarity of home. Desirous for change, they routinely set off on a journey away from their comfort zone.

The desire to leave the comfort zone is a common motivation for international students. The motivations for studying abroad in Doyle et al.'s (2010) work on Australian students included 'being put outside one's comfort zone' (p. 483). Chaban et al. (2011) have noted that one of the most cited reasons by New Zealanders to work overseas was for 'the challenge of living outside one's comfort zone' (p. 787). Talk of leaving the comfort zone is echoed in students' return narratives. Forsey et al. (2012) have indicated that upon return to their native Australia, students highlighted 'getting out' of their comfort zone as a valuable lesson gained from their studies abroad. Despite the ubiquity of the 'comfort zone' within travel discourses, the notion remains neglected within social and geographical scholarship. With this chapter I contribute a geographical focus on the 'comfort zone' by conceptualizing the notion through international student mobility.

I examine the notion of comfort zone through the mobility of Canadian students to the Global South. A total of 28 participants took part in the study. Among them 24 are women and 4 are men. The gender discrepancy in the sample reflects the over-representation of women in study abroad (Desoff, 2006; Schmidt, 2009; Stroud, 2010). Semi-structured interviews were used to explore the experience of Canadian participants studying or interning (from 3 to 12 months) in a developing country as part of a university exchange/mobility program (2012–2014). The interviews were conducted in three stages: pre-departure, mid-point, and return. Pre-departure interviews indicate that participants' motivation for leaving the comfort zone is to ultimately enable self-discovery and personal development/change. While this motivation merits in itself further examination and theoretical engagement/flexing, its purpose here is to contextualize the mid-point narratives which are the foci of this chapter. This chapter focuses on the interviews conducted during the sojourn abroad to demonstrate the interplay between participants varying zone(s) of comfort.

Although the comfort zone is a metaphorical term used in a variety of contexts within everyday parlance its definition remains relatively ambiguous. In its literal form, the term denotes a bounded, comfortable space. In common practice and usage, the comfort zone refers to a physical and mental space marked by feelings of security, familiarity and comfort. The physical space and material furnishings of a bedroom can constitute a comfortable space. Likewise, the presence of family members and friends can produce feelings of comfort. Similar to the multi-scalarity of 'home', the comfort zone can be projected at the scale of a house, neighbourhood, city, country or, as in the case of my participants, the Western world – although, the boundaries of the comfort zone are usually rescaled and reconfigured depending on spatial and emotional contexts. While the comfort zone can connote a positive space of comfort and security, from a different contextual perspective it can connote stasis, stagnation and lack of growth. Leaving one's comfort zone is viewed as a way to achieve self-growth, personal development and change (Brown, 2009). This is one of the main reasons that – at differing levels – people seek opportunities to move beyond their comfort zone. For young people, in particular, travelling abroad affords a way to 'step outside' of their comfort zone and challenge their personal status quo.

Social studies exploring the theoretical and conceptual boundaries of the comfort zone are scarce within the literature on travel. Brown's (2009) investigation of the practical applicability of the 'comfort zone model' in outdoor adventure education was the only study that emerged from my literature search that focused on the notion, albeit from a social psychology angle. The comfort zone model within adventure education is premised on the theory that placing people in challenging conditions involving stress and risk can result in individual growth (Brown, 2009). However, Brown (2009) suggests that overly stressful and risky situations can be counter-productive. Instead, such situations can induce anxiety that is detrimental to people's emotional well-being. Unnecessary risk or challenges may overwhelm and impede our ability to cope in stressful situations, and thus undermine any possibility of self-growth. In this chapter, I extend this perspective to international student mobility.

Although I assert that stepping into unfamiliar and challenging physical and emotional situations – and thus, stepping out of the comfort – is a potentially effective means to promote personal growth, I suggest that the stress of living abroad in an unfamiliar environment should be counterbalanced with periodic access to zones of comfort. I demonstrate how students seek out zones of comfort while abroad in order to recreate familiar feelings of 'home'. I argue that these spatial, corporeal and social zones alleviate feelings of homesickness and therefore enable students to regain the (emotional and physical) strength and urge to re-venture outside their comfort zone. Allen (2008) has indicated that his informants defined home in a variety of ways, which included 'feelings of comfort, familiarity, and safety', 'meaningful people and things' and 'intimate social and spatial knowledge rooted in place' (p. 89). The next sections discuss how participants recreate these three zones of comfort – spatial, corporeal and social – during the sojourn abroad. The spatial zone is configured within familiar, comfortable and homely locales and spaces while the corporeal zone consists of embodied consumptions of 'home/comfort' foods. The social zone involves the presence of, and (online) connections with, family members and friends. However, the boundaries of these 'zones' can converge both in theory and practice. For example, restaurants and other food-related places often include social and corporeal zones of comfort. As such, there is evident overlap between the three thematic sections.

The spatial zone

> It's funny how much of a bubble I managed to create. Sometimes I feel bad, like I'm cheating or not getting the right experience or something, because I still have access to internet, I still talk to people at home all the time. Again, I've managed to find a lot of food that I like; vegetables though not so much. So I've recreated a lot of that, so that's all been very much the same, which I feel bad about and at the same time, it's nice.
>
> [Alex, Canadian student interning in Rwanda]

The 'bubble' Alex refers to bears a resemblance to the notion of comfort zone. Indeed, through consuming familiar foods and internet contact with family she is recreating certain comforting conditions and homely qualities. In doing so, she expresses embarrassment and disappointment that she is failing to achieve the 'right' kind of experience. In other words, Alex voices concern that by recreating a comfort zone she is undermining her initial goal of the sojourn; that is, to achieve self-growth. Yet, she acknowledges that these zones of comfort afford positive feelings. Indeed, most of my participants admit that, at different moments in their sojourn, they recreate and seek out such zones of comfort. The term 'zone' is of course, by definition, inherently spatial. While I demonstrate in subsequent sections of this chapter that these 'zones' of comfort are also created through the social and corporeal, this section focuses on participants' spatial configuration.

Ahmed (1999) has commented that 'the unfamiliarity of a space/place can produce an experience of discomfort, or discomforting feelings' (p. 90). As international students settle in a foreign place, the spatial transition may be marked by a period of discomfort. Moving away from a familiar space or place can evoke a longing for the comforts of home. Case (1996) has noted that mobility attributes a sense of comfort to the meaning of 'home'. Spaces and places become 'home' when they are imbued with meaning and feelings (Easthope, 2004). In the case of mobile students, Hinton (2011) has argued that students carry and recreate feelings of 'home' in a new place. In recreating feelings of familiarity and comfort, Germann Molz (2008) has suggested that 'for some travelers, this sense of hominess constitutes a kind of "comfort zone"' (p. 333). In this study, international students associate spatial 'zones' of comfort with a sense of 'home'. Joan worked as an intern at a local

NGO in a village in Tanzania. During the mid-point of her 12-month internship placement I asked her what were her favourite places. She replied:

> The house of a friend. It's three people living together, and I met one of them and became friends with the entire house, and it's fun to go over there because I go run with them or use their oven to bake muffins or go over for drinks. I guess I just feel at home there and super comfortable like I could do anything. I don't need to worry about what I'm doing. I guess in those places I just feel at home.

Over time, specific places become familiar and comfortable. Hence, social spaces of comfort such as her friend's house evoke feelings of homeliness. Whether through the presence of friends or the comfy quarters of the house, these places represent a comforting space and thus a zone of comfort where Joan can retreat to feel 'at home' within a largely unfamiliar environment. This reflects recent work by Anderson (2012) in which the women in her study hinted to how 'the material and social characteristics of their living environments precluded or promoted a sense of belonging or comfort in New Zealand' (p. 331). Likewise, Case (1996) has pointed out that 'the simultaneous act of being at a familiar place and doing familiar activities with familiar objects is what evokes the sense of "being-at-home"' (p. 12). Thus, for Joan, the private sphere of a friend's place offers a sense of comfort and homeliness from the relative unfamiliarity beyond the doors of the house.

However, public spaces too can exhibit familiar and homely qualities. Christina, for example, has found public and commercial places during her year-long study exchange in Tanzania that resemble familiar places in Canada:

> So one of the first things that I realized that really made it home is when I found things, maybe not entirely like my home in Canada, but kind of as close as I can get here, like there's this one coffee shop that I go to here when I need to feel completely at home in Tanzania.... it just feels like a really cool Starbucks at home, which they even have cheese croissants which just sends me over the edge, it's amazing. So I've recreated the things that I would normally do but in a totally different way.

For Christina, the familiar features of the coffee shop, which reflect a popular North American coffee company, offer a space for her to recreate a sense of 'home'. Indeed, Collins (2008) has pointed out that 'in the case of Starbucks, the "third place" design of the now ubiquitous café franchise intentionally replicates iconographic environments that will be familiar to patrons regardless of whether they cross a city or a national border' (p. 164). Many participants have singled out local coffee shops in their host cities as favourite spots to recreate a sense of comfort and home while abroad. These spaces are thus appropriated to recreate a zone of comfort. Collins

(2008) further elucidates the popularity of such places among international students:

> many students identified in interviews that they regularly patronised ostensibly global franchises like Starbucks and other coffee outlets, and to a lesser extent places like McDonald's and Burger King, in ways that suggest these were also engagements with familiar senses and sociality.
>
> (p. 163)

In her study of global travellers, Germann Molz (2005) has noted that her participants talked about McDonald's as 'a site of familiarity and comfort' (p. 66) and a 'soothingly homey space' (p. 69). She describes McDonald's as

> a 'comfort zone' where travellers can indulge in 'comfort food' to ease the disorientation or difficulty of constant travel through foreign cultures. Travellers equate McDonald's to home implicitly by describing it as a space of familiarity, predictability, cleanliness, orderliness and belonging.
>
> (2005, p. 70)

Later in my interview with Christina she continued to identify local places and spaces that impart homely and familiar characteristics.

> There's this restaurant called Flirtease and I really love the restaurant because it has Wi-Fi and has really good French fries, and because it's so familiar and it's really comfortable. It has squishy seats and you can take off your shoes and get work done on your computer … so going to these places, like, I go downtown to this place, it's called Mokacity, and it's my favourite place downtown and they have ice cappuccinos, and it resembles Canada. So it feels really good, but in a totally different way. I look out the window and I'm completely reminded that I'm Tanzania, but I'm recreating what would be my safety net or what I would be doing on a Sunday afternoon.

Christina demonstrates how she uses familiar local spaces in Dar Es Salaam to not only recreate a sense of Canadian everydayness but also a 'safety net'; in other words, a zone of comfort. The familiar and comfortable spaces offered in certain cafés and restaurants evoked similarities with everyday spaces and places in Canada and thus, with 'home'. Indeed, Wiles (2008) has noted that a sense of 'home' was tied to familiar places and 'local ways of doing things' (p. 134). Participants reproduce a zone of comfort by spending time in familiar places and (re)creating everyday routines in these places. Christina also highlights particular food and beverages that make these places appealing to her sense of comfort and home. As Collins (2008) has observed, 'it is these embodied and sensual aspects of the restaurant that play such a crucial role in producing it as a 'familiar place" (p. 158). This spatial sense of familiarity and comfort is what enables participants to feel at home in their new environment. Physical features of specific private, public and commercial places exude certain qualities that are reminders of the embodied and material comforts of home. I suggest that participants appropriate familiar places within their host cities as comforting homely spaces. I argue that they frequent and use these spaces to recreate a zone of comfort and thus, a sense of home. Following on Christina's comments on food and the embodied comforts of restaurants and café, the next section examines the role of familiar and homely foods in creating a zone of comfort.

The corporeal zone

With the aim to 'get out' of their comfort zone, participants were keen to taste new dishes while abroad. During their sojourn, they experiment with different local exotic cuisines; they savoured the savoury *salgados* in Brazil, indulged in the delectable sweetness of *lokum* in Turkey and fêted and feasted at *asados* in Argentina and Chile. In Falconer's (2013) study of food tourist, backpackers experimented with strange foods and local eating practices that challenged their bodies and 'pushed their comfort zone' (p. 26–27). Duruz (2010) has highlighted 'the power of food to cross boundaries' (p. 47). Strange foods were consumed by participants as a way to remove themselves from their comfort zone. In her study of culinary tourists, Germann Molz (2007) suggested that 'instead of eating to feel at home, [they] eat to feel displaced, voluntarily engaging in the sense of outsiderness or alienation that can result from eating strange and unfamiliar foods' (p. 82). So while the novel assortment of flavours sampled by participants expanded their palate and imagination to new places, it also carried them further away from 'home'.

Bell and Valentine (1997) have discussed the role of food in memories of home. Food is central to reproducing familiarity and home (Bell and Valentine, 1997; Petridou, 2001; Collins, 2008). Traditional dishes can represent the nation and generate a sense of national identity and belonging (Rabikowska, 2010). Participants in my study show how familiar foods are intimately associated and tied to a sense of home. Marie-Anne described comfort foods from Québec and in particular maple syrup as 'a cultural reference point (*un repère culturel*)'. Returning to Christina's description of her homely spaces in Tanzania, she explains that her sense of home is 'often time associated with food. I guess, ya, this sense of home really connects with food and if I'm enjoying the food that I'm having.' She elaborates on the homely qualities of food:

> I'll come back and pick up some fruits through this crazy local market on my way back. Usually one week I buy the luxury of a block of cheese, which sounds pretty normal (laughs). (…) I'll buy a loaf of bread and

> sometimes even sliced ham, which wow! (laughs) which I manage for two days for lunch and its really nice; that's a really big homey feeling. So it's just recreating the things here in a different way that I appreciate about home.

Particular dishes have the symbolic and embodied capacity to recreate homely feelings. Eating familiar and enjoyable foods can draw participants closer to 'home'. For example, during her internship in Rwanda, Jane was one of many participants who described her food-centric attitude towards 'home':

> It's weird, I hate to say, but I'm very food motivated I think. You know, going somewhere where I can have a salad and nice coffee is such a luxury here (...) just going for coffee and a salad is hugely soothing because, I don't know, you feel healthy, it's something you would do at home.

Food is not only a source of nourishment and sustenance but also of source of comfort, familiarity and nostalgia (Locher et al., 2005; Brown, 2009; Brown et al., 2010). The experience of selecting a desirable food item from the menu and consuming that meal can produce a sense of hominess from which participants extract both bodily and emotional nourishment and comfort. Thus, food practices are sensory, embodied and edible pathways to 'home'.

As participants consume and experiment with new foods during their sojourn, the flurry of local exotic tastes can overwhelm and burden the senses and incur both physical and emotional discomfort. Whether it was the overabundance of spicy dishes, the bland taste of local food or the repetition and lack of variety of meals, a change in diet can take its toll on the body. Participants' well-being while abroad was influenced in part by the availability of certain familiar foods. The topic of food was most frequently raised when discussing bouts and feelings of homesickness. I spoke with Juliana who was half-way through her three-month internship in Peru. When I asked her what she most missed from Canada she did not hesitate:

> Peanut butter. They have it here but it's really expensive so I never want to go buy it, but I had peanut butter in a container with crackers when I came and that was sort of my staple snack for the longest time, because it's protein, it was tasty and it was home. (...) So when I was upset I would just sit there and eat this peanut butter with a spoon.

Although participants enjoyed sampling new local dishes, they sometimes sought out familiar 'comfort foods' to fulfil an emotional and embodied need for comfort and reassurance. Scholars have acknowledged that international students seek out and consume familiar foods while abroad in order to alleviate feelings of homesickness (Locher et al. 2005; Brown, 2009;

Abdullah, 2010; Brown et al. 2010). In a study on the psychological adjustment of international exchange students, Furukawa (1997) found that food had the greatest influence on intercultural adjustment. I argue that participants consume familiar foods to create a corporeal zone of comfort and thus, feelings of homeliness. As Locher et al. (2005) explain,

> Comfort foods were consumed at those times when the participants were feeling down and needed comfort, or when they needed an extra boost to get them through some task. Thus, comfort foods were perceived as special occasion foods only to be consumed when feeling blue or needing an emotional pick up.
>
> (p. 279–80)

This was the case of Brianne during her internship in Rwanda:

> you know when you're sick you have really specific foods you want to eat and don't want to eat and it was just like, 'I just want to eat Kraft dinner right now' (laughs); I just want a hamburger (laughs). Like, you just want really big American not-good-for-you food.

Émilie who is from Québec and is studying in Trinidad for a semester echoes this feeling: 'At the food court there is a KFC, Subway, Mario's Pizza and I go for a walk in the street and there is a Burger King, you know. So in any case, I'm really not [home]sick because of food.' Participants therefore have recourse to edible and emotional resources during periods of malaise and homesickness to support their continued participation in the sojourn and immersion outside of their comfort zone. Germann Molz (2005) has noted that participants in her study visited McDonald's during their travels 'to get a taste of home' (p. 64). As mentioned in the previous section, fast food joints such as McDonald's are a 'comfort zone' and a source of 'comfort food' (Germann Molz, 2005).

Despite the desire to step out of the comfort zone and partake in local culinary practices and exotic dishes, participants sometimes indulged in recognizable foods from Canada. Comfort foods are familiar and satiable; they become an emotional, sensory and imaginary repatriation to 'home' through their consumption. They stimulate and simulate a sense of hominess as well as a corporeal zone of comfort. Comfort foods are ingested to re-enter the comfort zone. However, I suggest that once participants have satisfied their appetite for 'home', they are then able to re-exit the corporeal zone of comfort. Thus, I argue that they periodically consume familiar homely foods to regain the strength and desire to re-engage with the unfamiliar and venture outside their comfort zone.

Participants' consumption practices interchanged between local culinary practices and comfort foods. They enjoyed eating new cuisine but occasionally sought out familiar foods too as a temporary respite from the unfamiliarity and discomforts of a new place. Different cuisines were consumed interchangeably to satisfy contrasting desires; that is, familiarity and novelty; comfort and exoticness. The longer the sojourn, the more frequently familiar foods were sought out, particularly during bouts of homesickness. Depending on time and their physical and emotional states, participants alternated between different edible geographies. Therefore, comfort foods represented an embodied, ingestible source of 'home' and a corporeal zone of comfort. The next section explores the role of family and friends in creating a social zone of comfort.

The social zone

Social relationships are among the prominent signifiers of 'home'. The literature is replete with studies that link geographies of 'home' to familial and social networks (Rykwert, 1991; Cuba and Hummon, 1993; Case, 1996; Mallett, 2004; Blunt and Dowling, 2006; Nowicka, 2007; Allen, 2008; Wiles, 2008). Scholars have echoed the adage of 'home is where the heart is' (Ralph, 2009). In their study of migration and a sense of home across the life cycle, Cuba and Hummon (1993) observed that the sense of 'home' for young people was primarily attached to social networks. Likewise, in examining a sense of home among New Zealanders in London, Wiles (2008) found that feelings of 'home' were associated to family. 'Home' is intimately tied to social relationships and this chapter expands on this conceptualization by arguing that connections with family and friends create a social zone of comfort.

In addition to craving comfort foods, periods of illness and homesickness elicited a longing for family in Canada. Brianne had suffered some health problems during her internship in Rwanda which made her realize the importance of family in her ideas and feelings of 'home': 'Being that sick made me really miss my family and want to be at home. (...) So ya, I definitely miss just being around people that I'm comfortable with.' Health problems highlighted the value of family in her feelings of comfort and home, and conversely, the absence of family exacerbated her feelings of homesickness. Family was an ultimate source of comfort.

As students bid farewell to their families and friends in Canada, they ventured forth into a new and unfamiliar social and cultural environment where encounters with locals and other internationals enable the creation of new social networks and relationships that can represent 'home'. Indeed, the creation of a new community is part of the home-making process (Castles and Davidson, 2000). During her year-long study exchange in Brazil, Véronique developed friendships with locals and international students that created a sense of 'home'. As she states: 'I think what makes home "home" is the people that you know.' In Kelletta and Moore's (2003) study on homeless young people staying in youth hostels, they noted that the social dimension of the hostel was central to their idea and sense of 'home'. While studying in India, Rachel received visits from her parents:

> I've had a number of close friends come to visit me in India and my dad is currently visiting me right now, so my dad is in the same town I'm in right now and my mother is joining us in one week so I have some of my very close family here. (...) So even though Canada still feels like home geographically, I have a lot of people that make home 'home' and they are a lot closer to me in India than they would be if I were in Canada (...). Having my dad here with me, there's this very obvious sense of familiarity and my mom is going to be here soon, so home is where I have a connection.

'Home', as Rachel illustrates, is embedded in the social. Rykwert (1991) has asserted that family *is* 'home'. Individuals are a mobile source of comfort and hominess through their travels and visits (Nowicka, 2007; Allen, 2008). As Wise (2000) has noted, people can feel at home 'simply in the presence of a significant other' (p. 299). People can move, and thus, 'home' travelled to and with her. Family members are thus a zone of comfort by their mere physical presence. In addition, Rachel forged new relationships with individuals abroad that also evinced feelings of hominess and comfort: 'I definitely felt like I have made strong connections in all the places I've been and because of those connections it gives me a sense of home and comfort.' Friendships developed during the sojourn overseas create a social zone of comfort and a sense of homeliness.

While the physical presence of family members and newfound friends during the sojourn abroad produce a social zone of comfort, similar feelings are also recreated through virtual presence. When participants connect to the internet, they are also connecting with family and friends. Through social media, participants can reconnect with the familiar presence of family and friends. Lillian describes her transition during her internship in rural China from attempting to adapt to the online social media restrictions to finally succeeding to gain access to Facebook so she could connect with her friends:

> There's this one point where I was like, I'm not going to use Facebook while I'm here, and I'm actually blocked by the firewall. But in the end I got a proxy to jump the firewall and was able to use it and I realize it's really hard for me to not keep up with what's going around with my friends outside, and that, I was really surprised, because I really want to just not use Facebook and be just here and in the moment. But I found that really difficult for me and I go on really more than I should or need to (laughs).

Lillian demonstrates her intention to be more than just physically present in China but also mentally and emotionally in order to achieve – what Jane termed earlier – the 'right' kind of experience. By connecting to Facebook, she is also emotionally and virtually connecting to 'home'. The internet and social media websites can represent a 'virtual home' (Germann Molz, 2008, p. 331). In other words, Lillian reveals the struggle between striving to remain outside of her comfort zone – that is, being disconnected from family and friends – and the desire to reconnect with her 'home' social network. Indeed, Germann Molz has pointed out that,

> Having an online presence and being beholden to a distant and dispersed audience of friends and family can be both comforting and oppressive. For some travelers, this level of connection and familiarity makes it difficult to actually leave home behind to achieve a sense of critical distance or personal growth.
>
> (p. 330)

Like many other participants, Lillian is torn between her motivations to escape her comfort zone and her desire for familiar and familial comforts. The mere act of logging in to popular social media sites can elicit comforting and homey feelings. I suggest that connecting *to* the internet and connecting *with* online social networks creates a 'virtual' (social) zone of comfort. Communicating and interacting with family and friends online can impart emotional support to participants to help them cope with moments of uneasiness and homesickness; which, left unaddressed (or in this case, disconnected), may lead them to retreat further into their comfort zone. I suggest that 'checking in' to such social media sites as Facebook and Skype establishes a temporary (re)connection to 'home' that provides them with the emotional recharge needed to virtually and mentally (re)disconnect from their comfort zone.

Conclusion

'Home', as I have shown, is intimately tied to comfort and familiarity. Although I acknowledge the existence of unfamiliarity and discomfort in geographies of home, I argue that for international students that voluntarily travel to study or work abroad, feelings of comfort were associated with feeling 'at home'. Comfort, home and mobility intersect and intermesh within participants' narratives of sojourns in the Global South. More specifically, 'home' and comfort are reconfigured through their international mobility. I have demonstrated how 'home' is projected at multiple scales, from the nation of Canada to local mundane places abroad, as well as manifested in different forms, from social relationships and comfort foods to virtual 'online' spaces. Participants feel secure and comfortable in familiar places and in the presence of friends and family, both physically and virtually. They also derived comfort from consuming familiar homely foods. Therefore, 'home' and comfort are tied to specific places, foods and to family and friends.

This chapter has showed that, despite participants' pre-sojourn motivations to leave their comfort zone, they seek out and recreate zones of comfort while abroad. I have argued that these zones of comfort are spatial, corporeal and social. The comfortable and familiar spaces of coffee shops and restaurants resembling those in Canada provided participants with a spatial zone of comfort. Overlapping with the spatial comforts of cafés and eateries are the edible features of these places. Consuming familiar and comforting foods created a taste of home and a corporeal zone of comfort. Most central to participants' feelings of 'home' were the presence of family and friends that created a social zone of comfort. Even by connecting to the internet and social media the 'online' presence of family and friends created a virtual zone of comfort. I have argued that such zones of comfort provide a temporary respite from the stresses and uncertainties of living, working and/or studying in a foreign country and allow participants to feel 'at home' while abroad. I suggest that these zones of comfort enable participants to regain the strength to re-venture outside their comfort zone and ultimately re-engage with the unfamiliar and local everyday life abroad.

In this chapter, I have attempted to disambiguate the conceptual and practical boundaries/parameters of the 'comfort zone'. In doing so, I have conceptualized the notion as a physical, embodied and affective space of familiarity and comfort that can be stretched, reconfigured, and recreated through mobility. Such a notion allows us to explore the emotional, physical and embodied resources used by international students in their home-making practices, and to understand their multi-scalar constructions of 'home' within mobile geographies and everyday life abroad. Finally, home-making practices and the zones of comfort help us to further understand the ebb and flow of international students' experience.

References

Abdullah, N. (2010). Comfort food, memory, and 'Home': Senses in transnational contexts. In D. Kalekin-Fishman and K. E. Y. Low (Eds.) *Everyday Life in Asia: Social Perspectives on the Senses*. Burlington, VT: Ashgate, pp. 157–176.

Allen, S. (2008). Finding home: Challenges faced by geographically mobile families. *Family Relations*, 57, pp. 84–99.

Ahmed, S. (1999). Home and away: Narratives of migration and estrangement. *International Journal of Cultural Studies*, 2, pp. 329–347.

Ahmed, S., Castañeda, C, Fortier, A-M, Sheller, M. (2003). Introduction. In S. Ahmed, C. Castañeda, A.-M. Fortier and M. Sheller (Eds.) *Uprootings/Regroundings: Questions of Home and Migration*. Oxford: Berg, pp. 1–19.

Anderson, V. R. (2012). 'Homes' and being 'at home' in New Zealand: Women's place-making in internationalised higher education. *Gender, Place and Culture*, 19(3), pp. 327–343.

Bell, D. and G. Valentine (1997). *Consuming Geographies: We Are Where We Eat*. London: Routledge, 248p.

Blunt, A. and Dowling, R. (2006). *Home*. London: Routledge. 304p.

Brickell, K. (2012). 'Mapping' and 'doing' critical geographies of home. *Progress in Human Geography*, 36(2), pp. 225–244.
Brown, L. (2009). The role of food in the adjustment journey of international students. In A. Lindgreen and M. Hingley (Eds.) *The New Cultures of Food: Marketing Opportunities from Ethnic, Religious and Cultural Diversity*. London: Gower, pp. 37–56.
Brown, L., Edwards, J. and Hartwell, H. (2010). A taste of the unfamiliar. Understanding the meanings attached to food by international postgraduate students in England. *Appetite*, 54, pp. 202–207.
Case, D. (1996). Contributions of journeys away to the definition of home: An empirical study of a dialectical process. *Journal of Environmental Psychology*. 16, pp. 1–15.
Castles, S. and Davidson, A. (2000). *Ethnic Mobilization and New Political Subjects. Citizenship and Migration: Globalization and the Politics of Belonging*. New York: Palgrave, pp. 129–155.
Chaban, N., Williams, A., Holland, M., Boyce, V. and Warner, F. (2011). Crossing cultures: Analysing the experiences of NZ returnees from the EU (UK vs. non-UK). *International Journal of Intercultural Relations*, 35, pp. 776–790.
Collins, F. L. (2008). Of kimchi and coffee: Globalisation, transnationalism and familiarity in culinary consumption. *Social and Cultural Geography*, 9(2), pp. 151–169.
Constable, N. (1999). At home but not at home: Filipina narratives of ambivalent returns. *Cultural Anthropology*, 13(2), pp. 203–228.
Cuba, L. and Hummon, D. M. (1993). Constructing a sense of home: Place affiliation and migration across the life cycle. *Sociological Forum*, 8(4), pp. 547–572.
Desoff, A. (2006). Who's not going abroad? *International Educator*, 15, pp. 20–27.
Doyle, S., Gendall, P., Meyer, L. H., Hoek, J., Tait, C., McKenzie, L. and Loorparg, A. (2010). An investigation of factors associated with student participation in study abroad. *Journal of Studies in International Education*, 14, pp. 471–490.
Duruz, J. (2010). Floating food: Eating 'Asia' in kitchens of the diaspora. *Emotion, Space and Society*, 3(1), pp. 45–49.
Easthope, H. (2004). A place called home. *Housing, Theory and Society*, 21, pp. 128–138.
Falconer, E. (2013). Transformations of the backpacking food tourist: Emotions and conflicts. *Tourist Studies*, 13(1), pp. 21–35.
Forsey, M., Broomhall, S. and Davis, J. (2012). Broadening the mind? Australian student reflections on the experience of overseas study. *Journal of Studies in International Education*, 16(2), pp. 128–139.
Fortier, A.-M. (2003). Making home: Queer migrations and motions of attachment. In S. Ahmed, C. Castañeda, A.-M. Fortier and M. Sheller (Eds.) *Uprootings/Regroundings: Questions of Home and Migration*. Oxford: Berg, pp. 115–135.
Furukawa, T. (1997). Cultural distance and its relationship to psychological adjustment of international exchange students. *Psychiatry and Clinical Neurosciences*, 51, pp. 87–91.
Germann Molz, J. (2005). Guilty pleasures of the golden arches: Mapping McDonald's in narratives of round-the-world travel. In J. Davidson, L. Bondi and M. Smith (Eds.) *Emotional Geographies*. Aldershot, UK: Ashgate, pp. 63–76.
Germann Molz, J. (2007). Eating difference: The cosmopolitan mobilities of culinary tourism. *Space and Culture*, 10(1), pp. 77–93.

Germann Molz, J. (2008). Home and mobility in narratives of round-the-world travel. *Space and Culture*, 11(4), pp. 325–342.

Hinton, D. (2011). 'Wales is my home': Higher education aspirations and student mobilities in Wales. *Children's Geographies*, 9(1), pp. 23–34.

Kelletta, P. and Moore, J. (2003). Routes to home: homelessness and home-making in contrasting societies. *Habitat International*, 27, pp. 123–141.

Kenyon, L. (1999). A home from home: Students' transitional experience of home. In T. Chapman and J. Hockey (Eds.) *Ideal Homes? Social Change and Domestic Life*. London: Routledge, pp. 84–95.

Locher, J., Yoels, W., Maurer, D. and Van Ells, J. (2005). Comfort foods: An exploratory journey into the social and emotional significance of food. *Food and Foodways*, 13, pp. 273–297.

Mallett, S. (2004). Understanding home: A critical review of the literature. *The Sociological Review*, 52(1), pp. 62–89.

McDowell, L. (1999). *Gender, Identity and Place: Understanding Feminist Geographies*. Minneapolis: University of Minnesota, 296p.

Nowicka, M. (2007). Mobile locations: Construction of home in a group of mobile transnational professionals. *Global Networks*, 7(1): 69–86.

Petridou, E. (2001). The taste of home. In D. Miller (Ed.) *Home Possessions: Material Cultural behind Closed Doors*. Oxford: Berg, pp. 87–104.

Rabikowska, M. (2010). The ritualisation of food, home and national identity among Polish migrants in London. *Social Identities*, 16(3), pp. 377–398.

Ralph, D. (2009). 'Home is where the heart is'? Understandings of 'home' among Irish-born return migrants from the United States. *Irish Studies Review*, 17(2), pp. 183–200.

Ralph, D. and Staeheli, L. A. (2011). Home and migration: Mobilities, belongings and identities. *Geography Compass*, 5, pp. 517–530.

Rykwert, J. (1991). House and home. *Social Research*, 58(1), pp. 51–62.

Skey, M. (2011). 'Thank god, I'm back!': (Re)defining the nation as a homely place in relation to journeys abroad. *Journal of Cultural Geography*, 28(2), pp. 233–252.

Schmidt, P. (2009). Men and women differ in how they decide to study abroad, study finds. *The Chronicle of Higher Education*. (Accessed online July 26, 2014) http://chronicle.com/article/MenWomen-Differ-in-How/49085/

Stroud, A. H. (2010). Who plans (not) to study abroad? An examination of U.S. student intent. *Journal of Studies in International Education*, 14, pp. 491–507.

Terkenli, T. S. (1995). Home as region. *Geographical Review*. 85(3), pp. 324–334.

Tuan, Y.-F. (1975). Place: An experiential perspective. *The Geographical Review*, 65(2), pp. 151–165.

Wiles, J. (2008). Sense of home in a transnational social space: New Zealanders in London. *Global Networks*, 8(1), pp. 116–137.

Wise, J. M. (2000). Home: Territory and identity. *Cultural Studies*, 14(2), pp. 295–310.

8 'Economia da Saudade'

Comfort food for London's Brazilian diaspora

Maria das Graças Brightwell[1]

Everyday food consumption is wrapped up in all sorts of desires and longings that go beyond our quest for nutritional requirements (Bondi et al., 2005). The very notion of 'comfort food', which has become almost ubiquitous, appearing regularly in popular magazines, television, literature and advertising in the last decade (Locher et al., 2005), brings together this entanglement between food, bodily desires and emotions. The *Oxford English Dictionary* defines this notion as 'food that comforts or affords solace; hence, any food (frequently with a high sugar or carbohydrate content) that is associated with childhood or with home cooking'. The social construction of some food objects as 'comfort food' must take into account both the physiological and social dimensions (Locher et al., 2005). There is evidence that sugar and starch make us feel good because they increase the level of serotonin in the brain (Christensen, 1977). Comfort foods also seem to guard a relationship with positive notions of nostalgia and sentimental values, with meaningful relationships and memories that guide people's choice of food in times of distress.

For migrants, the suggestion that consuming familiar food may provide the possibility of being transported to the 'whole world of home' (Sutton, 2001: 82), even if momentarily, and relieving oneself from the 'emotional overload' of living in a foreign land is a common feeling, as pointed out by Sutton. For immigrants navigating a foreign language and culture, a restaurant that provides familiar flavours is a city of refuge. (Kirschenbaum, 1939, cited by Jochnowitz, 2007: 131) Cooking and eating familiar food comes to the fore as home-making practices with high significance (Law, 2001, Petridou, 2001, Sutton, 2001, Marte, 2007, Rabikowska, 2010).

In this chapter, I examine the issue of comfort in relation to food and migration well beyond the ethnic undertones of 'comfort food' as an individual act of consumption to wider material, political and emotional aspects of migrant life and their presence in a host society. *Cafés*, restaurants and grocery shops owned and visited by immigrants play multiple roles in the immigrant's home-building experience. Recent works (Rabikowska and Burrel, 2009, Brightwell, 2012) suggest that these spaces are not only sites where business transactions occur. Rather, such commercial premises can act as important social spaces where immigrants socialize and seek solace and comfort in a strange land.

Of course, examples abound of host society ambivalence towards these spaces. As mentioned in the introduction of this book, comfort and discomfort permeate social encounters and constructions of difference. An example of host societies' ambivalence about migrant food consumption and the creation of migrant food commerce is brought by Wise (2011) in her study of the impacts of Chinese entrepreneurs to the urban landscape of Ashfield, Sydney. The production of a translocalised streetscape – in this case through Chinese language shop signage – 'impact upon the forms of belonging available to other inhabitants, sometimes causing tensions or competing claims of ownership' (Wise, 2011: 108).

Such distrust of migrants' decision to keep their food traditions can be rooted in a belief that ethnic food ties are a threat or an impediment to harmonic national identity. In America, for example, immigrants were encouraged (or forced) to abandon their traditional cultures in favour of adaptation in the host country (Kaplan et al., 1998: 124), occasionally going back to it in tougher economic times. In worst cases, immigrant food has been deemed unfit for consumption: Amy Bentley cites the folklorist Mario Montaño who notes that 'Anglos considered Mexican food so bad that they said wild animals did not scavenge on Mexican dead bodies' (Montaño, 1992 cited by Bentley, 1998). The literature also shows how food-related metaphors such as indigestion and contamination have been used to demonstrate fears about non-European immigration and racial intermingling in Australia (Edwards et al., 2000).

In this chapter I analyse the complex relationships that are found to incorporate food, migration and comfort as a means to consider belonging and home-making agencies and amongst the Brazilian diaspora in the UK. I follow Fortier (2005) in employing the notion of diaspora as a 'heuristic device' (Fortier, 2005) to think about questions of belonging, home and identity in the experience of dislocation among a relatively new migrant group. As pointed out by Fortier (2005) the new 'terrains of belonging' that are created in the new place of settlement have to re-articulate migrants' multiple locations, times, experiences and identifications. This intersectionality of belonging, which is not always accounted for (Valentine, 2007), should take into account that migrants' experiences in the new settings are mediated by their 'transnational histories, cosmopolitan attitudes, diasporic belongings, national identity, and particular positionalities of gender, race, ethnicity and citizenship' (Silvey and Lawson, 1999). Belonging is therefore, not given, but a process, entailing norms of affiliation (Fortier, 2000), emotional engagement (Hedetoft and Hjort, 2002) and a collective agreement to be created, performed and sustained (Bell, 1999). Belonging can also refer to feelings of 'being at home, in the sense of being surrounded by familiarity, safety and comfort, or the lack of these things, as when feeling homesick' (Hedetoft and Hjort, 2002: x). It is also important to note that Diasporic subjects negotiate their 'new geographies of identity' well beyond the realm of the nation states and across multiple sites of attachment (Lavie and Swedenburg, 1996). These multilocal mappings of 'home' imply, says Fortier (2005: 183) an examination of 'the social dynamics of roots and routings'. A number of important and challenging questions have been raised about home in the burgeoning body of migration, diaspora and transnational communities literature (Blunt and Dowling, 2006). Brah (1996: 193), for example, raises the point about the difference between 'feeling at home' and claims of belonging; Al-Ali and Koser (2002: 7) asks 'how transnational social fields and practices manifest themselves in daily lives, and how (if at all) do they impact on abstract conceptualizations of home?'

Through an investigation of food practices among Brazilians in London and its everyday spaces of commerce and consumption I bring together material, emotional, and political aspects of migrant life to the fore. By doing so I address the neglected emotional, embodied, material and quotidian practices in migrants' life (Halfacree and Boyle, 1993, Choo, 2004, Wise and Chapman, 2005, Conradson and McKay, 2007, Basu and Coleman, 2008). This body of work argues for the role of material culture in migrants' practices of belonging and home-making and to how 'diasporic identities are forged through the production, circulation and consumption of material things and spaces' (Crang, 2010: 139). Such work has addressed the relationship between migrant home-making and material culture through an engagement with architecture and gardens, furniture and personal belonging, and everyday activities and performances (Tolia-Kelly, 2004a, b, Morgan et al., 2005, Datta, 2008, Savaş, 2010).

I do this through an analysis of the pursuit of comfort by migrants, through the materiality of food and the spaces that its commerce engendered. I show how the setting up of importing systems by Brazilian entrepreneurs was central to the development of a stronger Brazilian food scene in London, as it allowed the expansion of the Brazilian grocery trade and the establishment of a more varied circuit of food provision – ranging from grocery shops, butchers, cafés and restaurants. Secondly, I argue that these retail outlets operate, not just as spaces of commerce but of sociality, providing emotional geographies of comfort and familiarity to displaced Brazilians. Thirdly, I bring evidence of the discomfort that food and its spaces of consumption bring: feeling trapped in one's own culture, providing visibility to Brazilian identity in London, something that can be both a positive affirmation but also a problematic identification of a population of which many members are 'undocumented' and wary of being visible to state authorities.

Methodology

The chapter is based on research for a PhD thesis completed in 2012, which investigated the role of food in the construction, representation and negotiation of Brazilian diasporic identities[2]. Part of the research involved a mapping of Brazilian food commerce in London, and subsequent visits to these commercial outlets, where face-to-face interviews with store owners, staff and clients, collections of menus, and observation and photo documentation were undertaken. In total 59 commercial establishments were visited. In each case, visits combined observational work and subsequent interviewing of outlet owners or managers. Overall, 30 participants who either worked in or owned Brazilian food outlets took part in the in-depth interviewing process. These individual, face-to-face, semi-structured interviews took place in the respondents' work environments. I also undertook focused periods of ethnographic research over the space of two months in two particular food outlets (one shop, one *café* / restaurant). Both were located in Harlesden, an area of northwest London now notable for its concentration of Brazilians. This work allowed me to investigate in more depth the social practices occurring in these spaces, and their role in the wider public culture of this neighbourhood. Although I did not use my own experiences as my only data source, I followed Pink's suggestion that 'the ethnographer learns about other people's experiences using her or his whole body and mind' (Pink, 2009: 1). This approach was very helpful, as I was trying to capture the creation and experience of these 'sensory landscapes of home'.

Using my own body and senses as a way of knowing about these places brought me intense emotions and feelings of comfort and discomfort. As a Brazilian migrant myself, I felt that these places triggered emotions and sensations, not always positive. In my field notes, I recorded being 'hot', bothered by the strong smell of meat, feeling full and lethargic after a meal, 'not feeling at home'; as well as feeling content, engaged and moved by the encounters I had, as this excerpt from fieldwork shows:

> As you enter the place, its laid-back style does remind you of a Brazilian grandmother's house, with its shabby and colourful plastic table covers and a multitude of objects hanging on the walls (at least my grandmother's house was like that!). As noted by the person who accompanied me when I went there to eat and by many other Brazilian customers who I spoke to, it is the sort of place that, despite its simplicity, you feel comfortable to present to non-Brazilians.

Further to this, I developed a parallel overview of Brazilian food consumption in London via four focus group discussions, which produced valuable and rich data of what were considered as being at stake in practices of Brazilian food consumption. Ethnographic work with Brazilian households, also located in Harlesden, complemented the data gathering. This work involved both observations and conversations focused on the role of food in Brazilian residents' domestic lives as well as, on occasion, testimony about their engagements with the commercial public culture of Brazilian food provision.

Economia da saudade: establishing networks of 'comfort' food provision

Almost invisible from public consciousness and academic studies, Brazilians form a significant part of the Latin American population in London and considered a 'manifestation of one of the "new migrant" groups that are contributing to the super diversity of the city' (McIlwaine, 2011). With this growth in the number of Brazilians in London, a niche market has also developed for specific products and services, which the immigrants have taken advantage of and have initiated entrepreneurial activities to respond to these new needs.

Brazilian entrepreneurs use the term '*economia da saudade*' (homesickness/nostalgia economy) to describe how a whole sector of Brazilian food commerce has been developed in London based on Brazilian migrants' demand of the 'comforting' foods and from home. 'Saudade', a deeply charged and 'key cultural word' (Silva, 2012) for Portuguese (and Brazilians as well) has been described as a 'deeply yearning for someone or something profoundly missed' (Duarte, 2005) and partly describes the complex feelings of some Brazilian migrants being 'out of place'.

Brazilian migrants play a fundamental role in both the provision and the consumption of Brazilian food. For instance, the setting up of importing systems by Brazilian entrepreneurs, previously dominated by Portuguese traders, was central to the development of a stronger Brazilian food scene in London (Aguiar, 2009). It allowed for the expansion of the Brazilian grocery trade and the establishment of a more varied circuit of food provision. Fuelled by a growing demand for Brazilian groceries from an increased number of Brazilian customers, the sector has grown more than tenfold in the last decade, from no more than 6 outlets to more than 70. Brazilian grocery traders in London, for the most part, have maintained stocks of Brazilian groceries and focused mainly on Brazilian clients. Brazilian food in London has a primarily diasporic rather than more generally globalised quality. Brazilian migrants are prime actors as both providers and consumers.

Apart from shops, restaurants, and cafés, the number of 'Brazilian' beauty salons, churches, magazines, and other community associations has also grown considerably since the end of 1990s when the flows of Brazilian migrants to the UK increased and peaked with the migratory restrictions imposed by the U.S. after 9/11. Precise numbers of migrants are not available and estimates vary greatly. The Brazilian government estimates that 180,000 Brazilians live in the UK and, similarly, community leaders and Brazilian researchers in the UK give the number of Brazilians living in London as somewhere between 150,000 and 200,000 (Evans et al., 2011, Evans et al., 2007), 30,000 of whom reside in the Borough of Brent alone. A recent research project on London's Latin American population shows a more modest number of 186,500 Latin Americans (including Brazilians) in the UK and 113,500 Latin Americans in London (McIlwaine, 2011). Despite such disparity in the figures, there is a consensus that Brazilian migration to the UK has increased since the 1990s and accelerated in the 2000s, albeit with a slight decline from 2007, due both to the recent economic recession and the tightening of immigration controls (Kubal et al., 2011, McIlwaine, 2011). Brazilians in London are a highly diverse group with regard to factors such as generation, gender, occupation, migration experience, social class, region of origin and religious affiliation. Studies have shown that despite a relatively high level of education, the majority lacks language skills and appropriate documents; they thus tend to find work in low skilled, low-paid service sector jobs such as cleaners, construction labourers, couriers and waiters.

Emotional geographies of comfort and familiarity

The importing of Brazilian brands by Brazilian entrepreneurs suffers restriction by import rules, therefore only a small variety of products and brands are sold. These are mass-marketed, mid-range products, sold throughout Brazil, and therefore present in most Brazilian kitchens: beans, chicken stock, coffee, soft drinks, seasoned flour, etc. They are, in Brazil, ubiquitous rather than special, mundane commercial products, more than national brands. However, these consumer objects acquire new meanings in the diasporic setting, namely the capacity to evoke familiarity for a large number of Brazilians, thus enabling the construction of collective spaces of belonging based on shared tastes and consumption practices. These brands became a way of giving form to diasporic culture and a means to 'transport' consumers 'home'. As Sutton (2001: 84) points out 'there is an imagined community implied in the act of eating food 'from home' while in exile, in the embodied knowledge that others are eating the same food'.

My intention here is not to argue that restaurants and grocery shops enable a collective identity project for all Brazilians in London. In fact, as became clear in this research's focus group discussions and domestic ethnography, there are Brazilians in London who do not identify with or go to these places. However, the ethnographic research in the *Brazilian food commerce in London* are illustrative of how food retail plays an active role in forging significant, visible, public spaces of belonging for some Brazilians in London. As such, they form community spaces that all Brazilians in various ways relate to, whether by visiting regularly, occasionally, or avoiding. The complex and new situations engendered by the lack of familiarity with London and the English language, often magnified by the absence of institutional and family support, mean that newly arrived migrants can suffer a great deal of emotional confusion. Familiarity brought by language, food, objects and social situations thus provides comfort and eases the burden. Brazilian food shops cater to such newly arrived Brazilian immigrants, those who have not learned the English language and cannot read labels or menus, identify products or order a meal in English. In that context exchanges are not limited to a monetary transaction nor simply the supply of material goods; they also comprise narratives about being Brazilian in London, reminiscences about their lives in Brazil, and the passing on of recipes, information about jobs, services, housing, education and immigration policing. What Brazilian food outlets lack in terms of sophistication and comfort is made up for by the imagination and cravings of a 'sensorially displaced' (Seremetakis, 1994) diasporic body that wanted a temporary fix of cultural familiarity. Personal feelings and problems are exchanged with total strangers, but strangers who could understand one's language and offer some comfort, or maybe even a practical solution.

Apart from the food itself, a number of other factors may contribute to the reconfiguration of a banal food shopping experience into a reminder of 'home' and 'homeland'. Mankekar (2002), in her fascinating analysis of Indian grocery stores in California, observed that the diasporic Indians who went to those shops were not only in search of specific products but interested in having an 'Indian shopping experience'. More generally, 'ethnic' restaurants and grocery stores rely heavily on embodied and sensual characteristics to attract customers and to make them feel at home. Collins (2008), in his study of the places patronised by South Korean international students in Auckland, describes some of these characteristics: (a) the presence of 'bodies' in these establishments – not only young Korean customers, but also the serving staff and cooks are Korean; (b) linguistic homogeneity; (c) the senses of smell, taste and touch provided by the food, the way it is cooked and presented, and the utensils provided; (d) a display of Korean objects on the walls, including masks, paintings, calendars, advertisements, drinks bottles, as a recourse to convey the 'Korean-ness of the place' and promote a 'sense of familiarity'. The notion of familiarity explored by Collins (2008:156) is thus experienced in 'a corporeal way to refer to sights, sounds, tastes, smells and touches that are known and as such stimulates feelings of comfort and belonging amongst those who experience them'.

The owners and staff of Brazilian businesses are well aware that, besides food, they need to provide a space where customers can respond to the emotional demands of migration and a diasporic life. In conversation with owners and staff, they are well aware of the emotional role that they and their place play:

> I feel very emotional when I talk about these things. This sixty-year-old person arrived here and stared at the Brazilian flag, maybe five minutes. A tear dropped from his eye. He was going through a lot of difficulty. He had not eaten for a while and he did not speak English. (Restaurant owner, 22/01/2010)

Having a place where people can go to eat 'mother's food', chat and reminisce, re-creates the possibility of 'making home' again. 'They take a little trip to Brazil whilst eating' says the owner of a café/restaurant and a market stall in Vauxhall and Brick Lane Market:

> It is true, people comment: 'Every Sunday my mother would make this manioc cake'. The other day, a customer commented: 'For me, my mum's chicken lasagne was the best in the world. I don't know if it is because I am far away, but yours is just like hers.' Food triggered in him a desire to go back, to be home. It makes a person remember.
> (Interview with coffee shop and market stall owner, 22/10/2009)

Such emotional geographies of displacement are an important part of how migrants experience daily life and construct their diasporic selves. Brazilian food places allow the sharing of these banal and localised forms of remembrance, provoked by smells and tastes: the kitchen, the table, the backyard, the family… By doing so, they may trigger a desire to go back. More importantly, through tastes and aromas Brazilian food commerce creates spaces where home can be re-lived and re-created under diasporic circumstances, which may assuage homesickness.

When home is not safe or comfortable

Of course, the sense of belonging provided by being in a familiar environment should not be assumed. One of the employees working at a Brazilian grocery shop reported feeling trapped within the cultural boundaries of Brazilian life in London. This resonates with similar feelings from respondents in the general research, some of whom were business owners who were frustrated by the fact that they were selling Brazilian culture when the objective of coming to the UK was to be immersed in the host culture. Migrants' themselves see this search for comfort food as limiting:

> I notice the way that memory and affect are important for [Brazilian] people who live here. They go out to buy manioc and chicken to make a dish that reminds them of their grandma and the taste of Brazil. […] It is something they miss… This emotion and homesickness makes them inward looking, they do not want to try out new things, they just want to follow the path they know, the path that is comfortable… (Interview with V., female)

Furthermore, a reminder of home does not always have positive connotations. The two cases I present here illustrate how painful the relationship between 'here and there' can be for immigrants. Familiar smells and tastes may take people back to their kitchen table but what happened there may bring painful memories of domestic and family life. These painful memories are shared with the staff in Brazilian commercial outlets. A., a Brazilian waiter told me that food is the trigger for good and bad memories:

> They tell me things that I'd better keep my mouth shut about. They talk a lot. They tell family stories. Some are pleasant, others not so. (…) You hear stories about that father who got drunk and violent.
>
> (Interview with A., waiter, 08/10/2009)

For many migrant workers, to go out and eat Brazilian food is a very special treat, one that perhaps in Brazil they could not afford. A., the owner of a Brazilian pizzeria, told me about the guilt felt by mothers who left their children in Brazil:

> Usually you get mothers thinking… most of them come here to earn money to help relatives, their families back home. Usually they say they get sad and they feel guilty because they are eating in a restaurant, spending money while their family is struggling back home and depending on them for financial help, so sometimes they feel guilty to be eating when they know, or don't know whether, their family back home is having a hard time.
>
> (Interview with A., owner, 08/10/2009)

Having seen how some mothers would buy a large quantity of biscuits and sweets, I asked the shopkeeper if this was a normal occurrence. According to her, Brazilian mothers 'fill' the children up with sweets and biscuits to compensate for absent and overworked parenting, as a substitute when a normal life with school, friends, relatives and open space was not possible:

> It is not given as a treat, but daily. I don't know if it is because I have been here for a while and I have changed my taste and my way of seeing things, but I think they give children too many sweets. And then they complain when the children put on weight or don't want to go to sleep until midnight. Maybe it is because in Brazil these treats are expensive and here they can afford them. Maybe they feel guilty because the children get bored because the parents don't do a lot with the children. Every day, after school, they come here and get a *coxinha* and a *guaraná* (soft drink). There is this client and she buys *Mucilon*, an infant cereal. The child is three years old and she buys a tin every three days. I asked the mother and she said that he doesn't eat anything else. The parents cannot say no to children when the subject is food. […] There is this emotional relationship with food: if the child is sad, they give chocolate to make it better. This happens a lot, because Brazilian children do not fit well in the schools. Many parents care more about making money than making sure their children are happy.
>
> (Fieldnotes, 15/06/2010)

Often, then, there was nothing celebratory in these doses of Brazilian diasporic culture, even if they came wrapped with memories of happier times.

Instead, they evidenced the marginalisation that migrant families, especially undocumented migrant families, experienced, and the temporary solutions they sought. The sweet taste of Brazilian diasporic food culture in this context was used as a palliative for a bitter and complex situation involving children, one that certainly deserves more academic attention.

Brazilian food commerce, I argue, both claims space for and gives visibility to Brazilian identity in London. Brazilian shops, in general, are one of the most visible signs of the Brazilian presence in London. With their overtly Brazilian façades they disseminate images and meanings of 'Brazilianess' for Brazilians and non-Brazilians alike. They create a 'banal' form of national identity in which 'Brazilianess' is literally 'flagged', as the national symbol sits in shop windows luring clients into commercial, social and cultural exchanges. This visibility via localisation was seen to be double-edged; both an assertive claim upon the space of the city; but also a risky intervention into a national space where legal presence was often denied. For undocumented Brazilians, the success of making Brazilian identity visible in Harlesden is not entirely welcome as it makes shops a target for the policing of migration status. At the same time as offering a place of sociability, these shops therefore also make such sociability a risky activity. For the shop owners too such visibility is double-edged: it is central to attracting custom, as well as being a source of personal pride; but it can also compromise the viability of their business, as their main consumers tend to shy away after raids. The visibility of Brazilians in Harlesden provoked immigration raids by the Home Office in 2008, which, for example, led to the deportation of staff and customers from both of my case study retail outlets. This has a negative impact on Brazilian businesses in the areas in general, at least for a while after, as customers either avoid going to these places altogether or come in and leave quickly. In April 2010, the immigration authorities came again to a Brazilian hairdresser next door to the grocery shop detaining and subsequently deporting a Brazilian woman. According to the shop staff, as a result, some people left the area, scared, and some even chose to go back to Brazil. Instead of being considered solely as a place of refuge or a positive affirmation of presence, Brazilian food shops in London can, sometimes, represent a place of danger and vulnerability.

Conclusion

This chapter addresses the discussion about the geographies of comfort/discomfort through everyday food consumption angle, showing how in the pursuit of solace and comfort migrants create spaces in London. In my ethnographic observations, these spaces of food consumption were sites where feelings of solitude, yearning, dislocation could be expressed and responded to; they were places where one could get in touch with the complex feelings and emotions that the migration experience brings. As we have seen, the emotional feelings of being 'out of place' and a desire to feel at home, to be comforted, were apparent for Brazilians across different social classes and regions, and with varying biographies of migration and settlement in London. Brazilian shops and restaurants, in this regard, provide the public locations where both the collective and private emotional aspects of identification and belonging can be played out. The notion of these retail spaces as part of an 'Economia da Saudade' illustrates perfectly the entanglement between feelings, materiality and place as people and objects move and settle.

Public spaces of food consumption such as *cafés*, grocery shops and restaurants not only reproduced 'sensory' experience of home through shared tastes, sounds and visual cues but also negotiate various notions of home via localised memories and particular tastes. However, two issues must be highlighted. First, the reminders of home triggered in these places were not always positive. Familiar tastes and smells may take people back to their homes; however, what happened there may bring painful memories of family life or may emphasise absence. Secondly, as pointed out by Brah, a distinction must be made between '"feeling at home" and staking claim to a place as one's own' (Brah, 1996: 193). To be able to call a place home and to belong to a place are defined by citizenship entitlement. The practices of home making and the creation of collective places of belonging in public places such food shops are intersected by fear, in a context where such visibility can be problematic for an 'undocumented' clientele. So the gaining of a sense of belonging and emplacement through food provision and consumption is not experienced the same way for all migrants. It is mediated by their national origins as well as by the national migration rules of the host land.

Such discussion about the comfort and discomfort engendered by migration processes – which entail encounters with difference, home-making agencies and dealing with complex feeling of belongings – could not be more timely, as millions of people leave their homelands in search of safety. We would like to conclude, by suggesting that seeking comfort is a human right, not a privilege of the few. The familiarity of food, in this sense, we argue, is one of the most enduring ways to provide the support and the strength, as pointed out in the introduction of this book, to satisfy the experience of everyday life.

Notes

1 CAPES Post-doctoral researcher at the Programa de Pós Graduação em Geografia, Universidade Federal de Santa Catarina, UFSC, Brazil.
2 Earlier versions of this chapter were published in Brightwell, M. D. G. S. L. 2012. 'On the move and in the making': Brazilian culinary cultures in London. *Canadian Journal of Latin American and Caribbean Studies,* 37, 51–80.

Bibliography

Aguiar, L. K. (2009) The expanding demand of Brazilian groceries: The case of cachaça. *In:* Lindgreen, A. (ed.) *The New Cultures of Food: Marketing*

Opportunities from Ethnic, Religious and Cultural Diversity. London: Gower, pp. 107–126.

Ahmed, S., Castañeda, C., Fortier, A.-M. & Sheller, M. (2003) Introduction: Uprootings/regroundings: Questions of home and migration. *In:* Ahmed, S., Castañeda, C., Fortier, A.-M. & Sheller, M. (eds.) *Uprootings/Regroundings: Questions of Home and Migration.* Oxford: Berg, pp. 1–22.

Al-Ali, N. & Koser, K. (eds.) (2002) *New Approaches to Migration: Transnational Communities and the Transformation of Home.* London and New York: Routledge.

Basu, P. & Coleman, S. (2008) Introduction: Migrant worlds, material cultures. *Mobilities,* 3, 313–330.

Bell, V. (1999) Performativity and belonging: An introduction. *Theory Culture Society,* 16, 1–9.

Bentley, A. (1998). From culinary other to mainstream American: Meaning and uses of Southwestern cuisine. *Southern Folklore, 55*(3), 238–252.

Blunt, A. & Dowling, R. (2006) *Home.* Oxon: Routledge.

Bondi, L., Davidson, J. & Smith, M. (2005) Introduction: Geography's 'emotional turn'. *In:* Davidson, J., Bondi, L. & Smith, M. (eds.) *Emotional Geographies.* Aldershot: Ashgate, pp. 1–18.

Brightwell, M. D. G. S. L. (2012) 'On the move and in the making': Brazilian culinary cultures in London. *Canadian Journal of Latin American and Caribbean Studies,* 37, 51–80.

Brah, A. (1996) *Cartographies of Diaspora: Contesting Identities.* London: Routledge.

Choo, S. (2004) Eating Satay Babi: Sensory perception of trasnational movement. *Journal of Intercultural Studies,* 25, 203–213.

Christensen, L. (1977) The effect of carbohydrates on affect. *Nutrition,* 13, 503–514.

Collins, F. L. (2008) Of *kimchi* and coffee: Globalisation, transnationalism and familiarity in culinary consumption. *Social and Cultural Geography,* 9, 151–169.

Conradson, D. & Latham, A. (2005) Transnational urbanism: Attending to everyday practices and mobilities. *Journal of Ethnic and Migration Studies,* 31, 227–233.

Conradson, D. & Mckay, D. (2007) Translocal subjectivities, mobility, connection, emotion. *Mobilities,* 2, 167–174.

Crang, P. (2010) Diasporas and material culture. *In:* Knott, K. & Mcloughghlin, S. (eds.) *Diasporas: Concepts, Intersections, Identities.* London and New York: Zed Books, pp. 139–144.

Datta, A. (2008) Building differences: Material geographies of home(s) among Polish builders in London. *Transactions of the Institute of British Geographers,* 33(4), 518–531.

Duarte, F. (2005) Living in 'the betweens': Diaspora consciousness formation and identity among Brazilians in Australia. *Journal of Intercultural Studies,* 26, 315–335.

Edwards, L., Occhipinti, S. & Ryan, S. (2000) Food and Immigration: The indigestion trope contests the sophistication narrative. *Journal of Intercultural Studies,* 21, 297–308.

Evans, Y., Tonhati, T., Dias, G. T., Brightwell, M. D. G., Sheringham, O., Souza, A. & Souza, C. (2011) *Por uma Vida Melhor: Brasileiras e Brasileiros em Londres, 2010.* London: Goldsmiths, University of London; Royal Holloway, University of London; Queen Mary, University of London.

Evans, Y., Wills, J., Datta, K., Herbert, J., Mcilwaine, C., May, J., Araújo, F. J. O. D., França, A. C. & França, A. P. (2007) Brazilians in London: A report for the

strangers into citizens campaign. 2007: Department of Geography, Queen Mary, University of London.
Fortier, A.-M. (2000) *Migrant Belongings: Memory, Space, Identity*. Oxford: Berg.
Fortier, A.-M. (2005) Diaspora. *In:* Atkinson, D., Jackson, P., Sibley, D. & Washbourne, N. (eds.) *Cultural Geography: A Critical Dictionary*. London: I.B Tauris.
Halfacree, K. H. & Boyle, P. J. (1993) The challenge facing migration research. *Progress in Human Geography,* 17, 333–348.
Hedetoft, U. & Hjort, M. (2002) Introduction. *In:* Hedetoft, U. & Hjort, M. (eds.) *The Postnational Self: Belonging and Identity*. Minneapolis: University of Minnesota, pp. vii–xxxii
Jochnowitz, E. (2007) From Khatchapuri to Gefilte Fish: Dining Out and Spectacle in Russian Jewish New York. *In:* Beriss, D. & Sutton, D. (eds.) *The Restaurants Book: Ethnographies of Where we Eat*. Oxford and New York: Berg, pp. 115–132.
Kaplan, A. R., Hoover, M. A. & Moore, W. B. (1998) Introduction: On ethnic foodways. *In:* Shortridge, J. R. & Shortridge, B. G. (eds.) *The Taste of American place: A Reader on Regional and Ethnic Foods*. Lanham, MD: Rowman & Littlefield, pp. 1–18.
Kubal, A., Bakewell, O. & De Haas, H. (2011). *The Evolution of Brazilian Migration to the UK*. Oxford: International Migration Institute, University of Oxford.
Lavie, S. & Swendburg, T. (eds) (1996) *Displacement, Diaspora, and Geographies of Identity*. Durham, NC: Duke University Press.
Law, L. (2001) Home cooking: Filipino women and geographies of the senses in Hong Kong. *Ecumene,* 8, 264–283.
Locher, J. L., Yoels, W. C., Maurer, D. & Van Ells, J. (2005) Comfort foods: An exploratory journey into the social and emotional significance of food. *Food and Foodways: Explorations in the History and Culture of Human Nourishment,* 13, 273–297.
Mankekar, P. (2002) 'India Shopping': Indian grocery stores and transnational configurations of belonging. *Ethnos: Journal of Anthropology,* 67, 75–97.
Marte, L. (2007) Foodmaps: Tracing boundaries of 'home' through food relations. *Food & Foodways: History & Culture of Human Nourishment,* 15, 261–289.
Mcilwaine, C. (2011) Super-diversity, multiculturalism, and integration: An overview of the Latin American Population in London, UK. *In:* McIlwaine, C. (ed.) *Cross-border Migration among Latin Americans*. New York: Palgrave McMillan, pp. 93–117.
Morgan, G., Rocha, C. & Poynting, S. (2005) Grafting cultures: Longing and belonging in immigrants' gardens and backyards in Fairfield. *Journal of Intercultural Studies,* 26, 93–105.
Petridou, E. (2001) The taste of home. *In:* Miller, D. (ed.) *Home Possessions: Material Culture Behind Closed Doors*. Oxford and New York: Berg, pp. 87–110.
Pink, S. (2009) Applied and participatory methods in visual ethnography, Intro duction to Visual Methods workshop, ESRC Research Development Initiative, Building Capacity in Visual Methods, University of Leicester 13–14 January.
Rabikowska, M. (2010) The ritualisation of food, home and national identity among Polish migrants in London. *Social Identities,* 16(3), 377–398.
Rabikowska, M. & Burrel, K. (2009) The material worlds of recent Polish migrants: transnationalism, food, shops and home. *In:* Burrel, K. (ed.) *Polish Migration to the UK in the 'new' European Union after 2004*. Farnham: Ashgate, pp. 211–232.
Savaş, Ö. (2010) The collective Turkish home in Vienna: Aesthetic narratives of migration and belonging. *Home Cultures,* 7, 313–340.

Seremetakis, C. N. (1994) *The Senses Still: Perception and Memory as Material Culture in Modernity.* Chicago, IL and London: The University of Chicago Press.
Silva, Z. B. (2012) Saudade a Key Portuguese Emotion, *Emotion Review,* 4, 203–211.
Silvey, R. & Lawson, V. (2004) Placing the Migrant. *Annals of the Association of American Geographers*, 89(1), 121–132.
Sutton, D. (2001) *Remembrance of Repasts: An Anthropology of Food and Memory.* Oxford and New York: Berg Publishers Ltd.
Tolia-Kelly, D. (2004a) Locating processes of identification: Studying the precipitates of re-memory through artefacts in the British Asian home. *Transactions of the Institute of British Geographers,* 29, 314–329.
Tolia-Kelly, D. (2004b) Materializing post-colonial geographies: Examining the textures of migration in the South Asian home. *Geoforum,* 35, 675–688.
Valentine, V. (2007) Theorizing and researching intersectionality: A challenge for feminist geography. *The Professional Geographer*, 59(1), 10–21.
Wise, A. (2011) 'You wouldn't know what's in there would you?' Homeliness and 'Foreign' signs in Ashfield, Sydney. *In:* Brickell, K. & Datta, A. (eds.) *Translocal Geographies: Spaces, Places, Connections.* Farham: Ashgate, pp. 93–108.
Wise, A. & Chapman, A. (2005) Introduction: Migration, affect and the senses. *Journal of Intercultural Studies,* 26, 1-1-3.

9 Assembling a great way to fly

Performances of comfort in the air

Weiqiang Lin

Introduction

Comfort has, in recent years, become once again a theme-tune of research in transport. While earlier work – or what one could label as 'traditional' transport studies (Shaw and Hesse, 2010) – has been preoccupied with the engineering of comfort as a means to enhance passenger experience during transit (see, for instance, Law and Cooperrider, 1974; Neveu et al., 1979; Teo et al., 2006), newer research has sought to steer this academic discussion towards a closer reading of its meanings, affects, politics and implications for mobility (Adey et al., 2012; Budd, 2011; Martin, 2011). This reorientation has led to a more serious engagement with comfort as more than a staid set of physiological concerns, related to how a general state of bodily ease can be achieved in transport. Instead, as Bissell (2008, p. 1697) contends, comfort is better understood as 'a highly complex sensibility... that requires sustained attention to the nuances therein.' Being 'comfortable', in other words, is never about a single logic or emotion to be stabilised while on the move; it is, rather, something contested, shifting and infused with particular ways of feeling.

This chapter joins recent efforts in re-defining comfort in such socio-culturally inflected ways, re-understanding it as an effect/affect that results from intersections between particular modes of producing and consuming it. But not wanting to rehearse previous writings on the ways human subjects react to, or whose corporeal sensibilities are complicit in, the formation of (dis)comforting registers (see Bissell, 2008, 2010, 2014), this research wants to interrogate how senses of comfort can be deliberately manufactured or 'worlded' as part of transport provision. This is not to suggest that current literature have not been aware of this point (see Jain, 2011; Watts, 2008 for some limited examples), but, in mobilities studies at least, there has been little sustained attention to the way vehicular spaces are being actively invented as dispensers and bearers of comfort, and not simply places for encountering or experiencing the same. Indeed, a keener appreciation of the minute organisational attempts at 'affective engineering' (Bissell, 2010, p. 280) onboard is potentially productive, not least in underscoring the strategic intents and politics woven into the bodies, materials and objects that make up transport even before it is utilised. Not to be mistaken as a privileging of structure over emergence, such a focus is able to cultivate an outlook that takes transport less as a ready-made space for occupation, than as a (pre)fabricated cultural phenomenon equally caught up in power.

In elaboration of this point, this chapter takes spaces onboard the aircraft as an exemplary site where passenger experiences are susceptible to being shaped and manipulated. Here, the travails of Singapore's flag carrier, Singapore Airlines (SIA), to induce an ultra-comforting in-flight atmosphere, by means of a self-concocted brand of Orientalism performed through flight attendants dubbed 'Singapore Girls',[1] will be investigated. The provision of a 'top-notched' cabin service by Singapore is not just a business gimmick, but an impetus that is inextricably linked to the small city-state's fortunes following the end of British colonialism. Without a sizable market to sustain its own traffic, and faced with new hub rivals in decolonising Southeast Asia, creating an airline known globally for its unparalleled comfort proves critical to Singapore's ability to buttress itself as a key transfer hub in the region. Remarkably, achieving such an effect does not (only) come with the use of superior 'hardware' such as more ergonomic seats, but draws heavily on the 'emotional labour' (Hochschild, 1983) of SIA's cabin crew, whose Orientalised 'body-work' helps set the carrier apart (Chan, 2000). By expounding on this strategy, I aim to draw attention to a less-researched genre of comfort production in transport, involving culturally inscribed bodies that, this time, are not positioned to seek corporeal ease for themselves, but rather give it.

The rest of this chapter is divided up as follows. In the next section, I will flesh out how this particular, 'non-Western' example in transport-space-making can contribute to scholarships on comfort and mobilities. Besides reviewing how authors have currently approached transit spaces as (micro) environments where (dis)comforts are being felt, negotiated, and struggled over, the section will also reflect on the distinctive cultures of the aircraft cabin, as a realm often fashioned by the industry to convey particular moods – of safety, light-heartedness, charm etc. – to passengers. After a brief note on SIA's history and methodology in the third section, I go on to interrogate, in the fourth section, how SIA has from the outset sought to carve out a competitive space for itself by introducing the iconic 'Singapore Girl' to its cabins, embodying in-flight space through a selective Orientalism that is at once alluring, familiar and efficient. Recognising this as a move to ensure the city-state's continued connectivity and survival, this chapter concludes with a few provocations on aeromobile futures in a world where aircraft cabins become instrumental to the production of aerial networks.

Taking comfort in transport

Recent research on mobilities has sought to more clearly define what transpires in the mundane but socially significant corridors of transport

(Binnie et al., 2007). Acknowledging the eventful-ness of vehicular spaces, scholars are stressing the range of human agencies and cultural practices enacted by people during their journeys. One aspect that has been animated pertains to the (dis)comforts of travel. Specifically, there has been increasing recognition of these emotions as something affectively negotiated and culturally implicated. As David Bissell (2008) observes, comfort is not just passively conveyed or transferred, but is the dialectical exchange between the body and its environs. It is *through* the body that one feels 'a myriad of affectual complexes and sensations' relating to (dis)comfort, shifting as 'comfort ruptures, splinters, and moves beyond what could be deemed a pleasurable sensation' (Bissell, 2008, p. 1709). Consequently, a range of different experiences – even frustrations – become possible while on the move, including not just the comforts that a vehicle is designed to afford (Jain, 2011), but also feelings of stress (Bissell, 2014), fatigue (Adey et al., 2012) and motion sickness (Ashmore, 2013) that arise as bodies move in and out of their desired states. In short, even in the midst of apparent sedentarism, the transported body does not stay still. Instead, it actively seeks to ensconce itself within and/or restore itself unto particular affective dimensions and relations with others.

If the above work is concerned with how the body attempts to find a stable, comfortable posture in a given situation, another thread examines how social interactions can likewise engender particular states of (dis)comfort through (non)relations with people. Underscoring the collective nature of public transit, Laura Watts (2008) notes how a shared sense of community emergent onboard a train can impel particular acts of civility – such as being quiet – among members, so as not to provoke others to sensorial discomfort. Also alluding to the 'art and craft' of travel, other authors draw lessons from the effects that 'unacceptable' behaviours can have on the mood of travel. Helen Wilson (2011, p. 639), for one, reveals how the bus can be a highly circumscribed space, replete with 'unspoken rules' on seat sharing and 'polite' ways of talking. While, in her study, the failure to adhere to these rules can be deemed irksome by some, Bissell's (2010) research on rail travel detects a similar sort of misanthropy among passengers when this usual calm is broken, bringing to light society's increasing intolerance of, and discomfort with, moving proximately with strangers. The mobile spaces of transport are, as such, sites that have far-reaching implications beyond their transient selves. Not only can the violation of transit norms instil a jarring sense of unease among travellers, it can also (re)shape societal views, through what takes place in the microcosmic spaces of transport.

In the same vein as buses and trains, the aeroplane arguably provides a comparable context for such registers of (dis)comfort to arise. For the most part, burgeoning work in this area has concentrated on the physically uncomfortable nature of air travel, not least in relations to the insalubrious cabin environment that passengers have to endure for increasingly long hours (Bissell, 2015). In fact, such discomforting – often intertwined with health – discourses have surrounded aviation for a long time, traceable to when balloon flights took off in the mid-1800s (Johnson, 1943; see also Budd, 2011; McCormack, 2008). Following World War I, the focus on healthful aerial science and more hospitable environments for pilots and passengers only grew (Gibson and Harrison, 2005; Simonsen, 2010), as more knowledge was gained in the field. Notwithstanding this, and despite the many modern-day advances in cabin design, 'life in the air' remains an unsettling one, characterised by maladies such as cramped seating, aviophobia, sub-optimal breathing and eating conditions, dehydration and jet-lag (Budd et al., 2011a, b; Gottdiener, 2001). Against earnest efforts to tame the hostile elements of the air, what predominates in the 'cocooned' interiors of aircraft (cf. Martin, 2011) are lingering signs of the medium through which aeroplanes traverse, with the 'outside' always threatening to upend the carefully calibrated climate (of restorative comfort) that plane-makers are trying to contain 'inside'.

Yet, air travel is also unlike most other modes of transport. Though cabin space can never be fully expunged of the physical discomforts of flying, it is often specially choreographed to be filled with alternative encounters designed to make one forget about the cause of that unease. While Bissell et al. (2012) point to the use of humour in in-flight safety videos to soften grim reminders about the dangers of flying, other scholars have reflected on 'the service function latent within these entubulations' (Adey and Lin, 2014, p. 67), which tries to counteract the ills of Icarian dwelling with a more reassuring 'normalcy' sustained by flight attendants. As Drew Whitelegg (2007, p. 35) notes, the profession of the 'stewardess', when introduced in 1930, had been key to delivering a positive 'psychological punch' among businessmen who were then discomfited by the idea of flying. Similarly in the 1960s, liberalising sexual attitudes saw airlines in America begin selling the sexual appeal of female crew as another strategy of appeasement, metamorphosing stewardesses from 'dowdy nurses' into wearers of 'micro-mini skirts and hot pants' (Hayes and Tiffney, 2007, p. 117); and surrounding them with risqué taglines, like National Airline's 'Fly Me' (Crang, 2002) and Pacific Southwest's 'Pure, Sober and Available' (Omelia and Waldock, 2006). Meant to placate (male) passengers, the aircraft cabin is thus always more than a static place of waiting (Bissell, 2007). Airlines and aircraft manufacturers do not only seek to ameliorate the discomforts of their occupation with material and technological innovations (see Govil, 2004 on in-flight entertainment), they also permeate them with service tropes meant to divert attention from their inadequacies as comfortable realms.

Merging these different threads on (dis)comforts, bodies, social interactions and cabin spaces, the rest of this chapter will consider how a specific airline – Singapore (SIA) – attempts to reduce the unease of flying by, likewise, assembling an alternative encounter for passengers. Beyond the installation of 'hardware' such as better seats and more luxurious accommodation, its flight attendants are critically called upon, too, to connect with passengers, and to construct for them a less harsh 'world' onboard. Such a tactical deployment of frontline staff invokes what Arlie Hochschild (1983, p. 7) termed 'emotional labour', a work that requires, in this case, cabin crew, 'to induce or suppress feeling', in order to instil in passengers a sense of being cared for. While this is nothing new in the airline industry, SIA's particular formulation is notable for its overt manufacture of 'comfort' as a means to differentiate itself in an industry that was, at the time of its inception, beginning to be catered to the masses (Boris, 2006). What is more, SIA's human-simulated comfort is one that is simultaneously imbricated with an invented cultural sensibility, characterised by 'Asian-ness', exotic difference, but yet familiarity. Before tracing this particular instance of comfort prefabrication in a vehicular space, I now turn to the airline's history to better understand its mission and context.

Singaporean departures

The history of SIA has often been traced to its colonial predecessor, Malayan Airways, a joint venture between Ocean Steamship Company, Straits Steamship Company and Mansfield Company in the 1940s (Raguraman, 1997). Following a labour strike in 1956 over income disparities between British expatriates and locals, the airline was gradually nationalised and eventually renamed Malaysia-Singapore Airlines (MSA) in 1965. In the late 1960s however, this joint venture was deemed unsatisfactory again, this time by the newly-independent city-state which felt that Malaysia was using MSA to grow its domestic network (Raguraman, 1997). At that time, the island-nation, devoid of natural resources and a viable economy (Chua, 1995), had wanted to ensure that it could continue accessing world markets through developing robust aerial links, even as the end of British colonialism had cut formal/former trade links. This was made more urgent by the rise of neighbouring aerial competition from the likes of Bangkok and Kuala Lumpur, which had wanted to depose Singapore as the main Southeast Asian hub (Office of the Commissioner-General, 1962). It was in this context that SIA was born in October 1972 under the auspices of the Ministry of Finance, to serve as the nation's international arm and vehicle and to help direct and buttress long-haul traffic – then, comprising mostly Western passengers – to the city-state.

Helmed by Chairman Joseph Pillay and Deputy Lim Chin Beng, SIA was envisioned to be an airline whose foremost 'commitment' was to the customer (Wirtz and Johnston, 2003). Its leaders had set out to recruit a cabin workforce that would make in-flight experience more enjoyable than even that found aboard Western veteran carriers. As an SIA commemorative issue writes,

> Emphasis was placed on servicing the passenger, who... was the reason for SIA's existence. The cabin crew were the prime link between

> passenger and the airline: in a world where one airline looked much like another, their task was to make sure Singapore's national carrier was remembered.
>
> (Allen, 1990, p. 69)

More than simply restoring industry service standards, SIA was really looking to compete internationally and to be 'remembered' as *Singapore's* national carrier. This required its prime space of business, the cabin, to be properly differentiated from the rest, and be marked with a hospitality that would carry 'the indefinable magic of the East into European hearts' (Allen, 1990, pp. 53–54). At the centre of these enchantments was the Singapore Girl, an embodiment of comfort that would soon become shorthand for good service. As the commemorative issue continues, she was to provide 'in-flight service [that] was better than anything anyone had experienced for a long time' (Allen, 1990, p. 72). Conceived to make (male) passengers forget about the realities of travel, she was purposed to help manufacture a new onboard experience that would redefine the meaning of in-flight comfort.

Before going deeper into this story, a brief methodological note is needed. Supplemented by relevant news articles, this chapter draws on archival research on two staff newsletters – *HighPoint*, for the cabin crew; and *Outlook*, for the entire company – to tease out the roles that Singapore Girls play in achieving this state of comfort. In particular, I focus on the period between 1972 (the company's founding) and the early-1990s, when the airline was beginning to be more preoccupied with market expansion than penetration. Admittedly, relying on company materials for evidence risks missing out on flight attendants' actual delivery, or how they might resist their scripts (Murphy, 1998). This chapter's objectives are as such calibrated to highlight only the *intended* outcomes of SIA's designs. Nevertheless, for an airline that has consistently received top honours for its service, and whose training programmes are reputed for their efficacy as a form of corporate disciplining (Chong, 2007), these newsletters must, to some extent, be recognised as more than propaganda, connoting some degree of actualisation of the 'excellent service' (Wirtz and Johnston, 2003) aspired to. The emotional labours examined are, in turn, not completely fictitious, but in some ways contribute to helping SIA become a top airline of choice among international travellers.

Assembling a great way to fly

Corporeal allure

As the first step in making journeys more enjoyable, the Singapore Girl had to personify that 'Asian' promise of care and comfort from the start. This was achieved through a careful sculpting of her corporeality, so that she would leave the appropriate impression on passengers. Yet, the Singapore

Girl was also a thoroughly Western conception. Starting with her uniform, French designer Pierre Balmain was invited to conceptualise a fitting dress for her, for which he chose the sarong kebaya – a traditional Malay blouse-skirt ensemble once worn by native women in colonial Singapore (Lim, 2000). The snug-fitting costume, matched by batik slippers, was meant to amplify her svelte figure and to bring out her comforting demeanour as a 'graceful Asian young lady [in service]' (*The Business Times*, 2007). In contrast to Western flight attendants who wore formal jackets and skirts, the Singapore Girl was thus to make her first appearance in an indigenous dress that suited a Western(er's) imagination of 'Asian' hospitality, already setting the tone for the kind of service she would deliver.

Yet, the charm that the sarong kebaya exuded would not have been as potent if not for a second round of touch-up offered by a British man. Ian Batey, a creative designer, was hired by SIA to define the Singapore Girl more sharply as a symbol of comfort. Having secured SIA as his first client, he set on the task of producing a simple soft-focused photograph of a Singapore Girl's profile, with the seductive caption 'This girl's in love with you', as the airline's first commercial print (Wee, 1999). By evoking exoticised and eroticised notions of the 'Asian' woman, the message was an intentionally gendered and Orientalised (Said, 1978) one, relating the Singapore Girl with her eagerness to please and administer loving care. As an accompanying jingle from the 1970s dreamily vocalised it, 'Sunlight in your smile, such a lovely style, stay with me a while – Singapore Girl. Everywhere I go, your smiling face goes. Floating through my mind – Singapore Girl, you're a great way to fly' (Hart and Bongusto, 1979). Framed by such thinly veiled colonialist innuendos about Asian gentleness and sexualised comfort, the Singapore Girl had become a symbol and visual spectacle that preceded even her labour, poised to change the tenor of SIA's cabins from one of arduousness to one of pampering by virtue of her titivating presence.

But tracing the Singapore Girl on paper was not enough; she had to be brought to life according to the image given her. In a sign of fastidiousness and preference for younger girls, the company would organise recruitment drives at local high schools to attract female students to the glamour of flying at an early age. Consider this 1985 entry in *Highpoint* (1985a, p. 20):

> Singapore Airlines had distributed 4,500 postcards to students who sat for their 'A' level examinations at the end of last year, inviting them to become flight stewardesses. The Company wants to attract these girls who are 18 years and above, as this is the minimum age for flight stewardess.

Young girls were invited to apply as Singapore Girls starting with those of the (regulatory) minimum age of 18. Then, in a gruelling sifting process whose success rate could be as low as 2 per cent (*Outlook*, 1986a), applicants would undergo rounds of selection and probation, which comprised – as it still applies today – interviews, group exercises, and, upon preliminary selection, stringent training programmes. One of the most notorious components in the training programme was that the girls had to take a swimming test, from which their male counterparts were exempted. While this was a crucial part of ensuring safety, a cabin personnel manager admitted in a Belgian documentary that this was conducted so that management could 'check out' and ascertain the women's corporeal suitability for the job (Whitelegg, 2007, p. 135). By excluding matronly figures or those with unseemly scars and blotchy complexions, SIA was then able to ensure that only pristine bodies could serve as its ambassadors of 'Asian' comfort.

The Singapore Girls were also directly enrolled to upkeep their appearance, through a raft of company rules prescribing how they should appear and present themselves. For make-up, they would have to adhere to a particular blend of colours, in order not to 'look ghastly under the orangey cabin lights' (Tan, 2000, p. 4). Reverting to Western expertise, a mix concocted by French make-up stylist, Olivier Echaudemaison, was selected:

> [t]he colour scheme for our stewardesses was specially created… to complement our batik uniform – deep blue, pearly pink, rust and pale gold for the eyes, peach blush for the cheek, and shiny Corail-red lip colour to 'bring out the life in that smile'.
>
> (*HighPoint*, 1983, p. 24)

Besides these foundational measures meant to conjure 'a complete sense of coordination between a woman's face and what she is wearing' (*HighPoint*, 1983, p. 24), tips on a variety of beauty topics were also conveyed to ensure that female crew would always exhibit an unfailing sense of immaculateness. These included: how to appropriately care for one's skin (*HighPoint*, 1986a); the efficacies of 'natural beauty aids' like turmeric powder, cucumbers and gingerly (olive) oil in returning a suppleness to one's hair, face, hands and legs (*HighPoint*, 1986b); and what 'slim' and 'desirable' range one should keep one's weight within to enhance 'your physical appearance, poise and confidence' (*HighPoint*, 1990a, p. 14). As former flight attendant, Phyllis Lee, explained, Singapore Girls were instilled with the gravity of looking good from the time of training, when '[w]e had a mirror in the classroom, and everyone would run into the bathroom to touch up during breaks' (Lee, 1993, p. 2). By inculcating the 'utmost importance' of vanity, SIA was effectively gaining its crew's complicity in maintaining the image it had so designed.

From senior management to the creative designers, there was clearly a concerted effort to sell to prospective passengers a visual experience that detracted from the usual discomforts of flying (Budd et al., 2011a). For SIA, passengers' comfort did not solely rely on the material affordances of the aircraft, but must begin with the rhetorical assurance that, inscribed upon the impeccable faces and bodies of its flight stewardesses was a sign of 'Asian hospitality' that no other airline could replicate. Such service pledges, albeit grossly sexist, were arguably aimed at altering the tenor of the cabin, and rendering SIA flights a mobile space to be desired. In a meeting between the (Western, male) passenger's gaze and the airline's vision of *who* could deliver 'good service', the Singapore Girl's corporeal allure was, by design, meant to serve as a pretext for passengers to react to once onboard. Her delectable presence, drawn from a repertoire of old and new, Western and self-Orientalising scripts, not only promoted the airline's viability in a world of giants, but also kick-started Singapore's bids to guide the world's traffic to its fold.

Familiar demeanour

To further reconcile the ideal Singapore Girl with her 'real' self required more than just aesthetic appearances. In an industry where whiteness, including that of flight attendants (Barry, 2007, p. 12), is often equated with technological aptitude, the ability to lay claim on Oriental-ness must be tempered with a conduct that would avoid making SIA's 'Asian' flight attendants seem technologically inept by default. Accordingly, the airline had taken steps to try to preserve, through its cabin crew, a sense of calming comfort and (thus) trustworthiness in the aircraft cabin wherever possible. Expressly, it had conscientiously trained its crew in the way that they should approach passengers, without provoking them to undue stress, emotional disturbance or unease about the airline. To achieve this, the Singapore Girls had to acquire new, and sometimes alien, dispositions counter-intuitive to them, but that were familiar to their (Western) audience. Contradictory to earlier depictions of her natural 'Asian' charms, she was now to be rehabilitated in the ways of 'proper' hospitality, and have her tropical habits reined in.

One of the first 'corrections' that SIA's cabin crew had to undergo was to lose their 'unpleasant-sounding' Singaporean accent. Beginning 1984, British Council instructors, equipped with custom-made English language coaching materials, were employed to offer them 'speech therapy' and help 'improve the quality of in-flight announcements [among] crew members who [got] a B grade for their announcing ability' (*HighPoint*, 1984, p. 19). Students had to attend two four-and-one-half hour tuition sessions, before shifting onto individual practice with a private tutor, who would give personal advice on how to improve, and to make impromptu announcements fluently. Through these lessons, crew members were expected to master 'how [to] use your voice to give a favourable impression of yourself and the airline you represent', and

> use 'referring tones', which sound more friendly than the 'proclaiming tone' (*HighPoint*, 1984, p. 19). A Cabin Crew Training Centre completed in 1979 additionally availed to flight attendants a 'language laboratory', where they could access a wide selection of coaching tapes 'to polish their linguistic ability'.
>
> (*Outlook*, 1980, p. 8)

Presumably, their natural speech – which one *HighPoint* (1990b, p. 4) issue referred to as 'polluted English' – would tarnish the good name of SIA, communicating a jarring image antithetical to the airline's image.

In another attempt to put passengers at ease, the airline found it imperative to inculcate crew members with the 'right' values of social etiquette. SIA's consultant, Jennie Nicholas, an Australian, was placed in charge of this task. At her training school, flight attendants learned the ways of 'quiet politeness'. As a journalist who visited the campus reported, the school's entrance was adorned with a banner that read 'We depend on the customer, not they on us', while 'classes began with the girls repeating: "A passenger is human, with biases, feelings and emotions like our own"' (*The Sunday Mail*, 1986), which were best left unperturbed. Indicative of whom these customers that SIA was trying to appease might be, every stewardess had to be trained in the manners of Western fine dining, even as chopsticks were banned in the school. An entry in a *HighPoint* issue (1986c, p. 3) alluded to this when it featured a story of Nicholas's etiquette course on 'the continental style of table manners and the proper use of cutlery'. Taboos such as 'coughing over others; sneezing without placing the hand over the mouth; combing hair in public' were further raised as gestures to avoid. Why would anyone behave otherwise? The answer was 'very simple', according to *HighPoint* (1986c, p. 3): 'few of us really have the opportunity of getting the correct information regarding social etiquette… [but it would] not [be] too late for people who had already chalked up many years of socially unacceptable behaviour to change'. It was as if, for all the desirability of an Oriental crew, it now problematically came attached with latent defects and cultural inferiorities, which needed to be purged from an already-nerving cabin space.

If the above behavioural corrections ensured that the crew comported themselves in ways that synced with passengers' ideas of a familiar, and thus trustworthy, airline, another series of programmes sought to persuade them to adopt attitudes that lived up to 'global' standards. In particular, a prominent movement in SIA in the 1980s had been to instil, among flight attendants, 'courtesy' as a way of life. This agenda had roots in a wider campaign in the city-state at that time to inspire citizens to be more considerate, and to create a pleasant social environment for residents, tourists and investors (Tham, 1983). As the global face of her country, the Singapore Girl naturally became a starting point for SIA to enjoin itself to this effort. Specifically, SIA's courtesy campaign would bring Singapore Girls' aerial performances to terrestrial grounds for rehearsal. Since 1982, 'cabin crew courtesy divisional cells' and a 'courtesy committee' were set up for staff to mingle in, have meals together, and practise 'acts of kindness' with each other (*HighPoint*, 1986d, p. 5). By encouraging staff to establish closer working relations, and to cite both courteous and discourteous acts among themselves, the airline's hope was to nurture that desired 'someone who shows concern and respect for his *[sic]* superiors, peers, subordinates and the public through his/her manners, behaviour, polite speech and patient listening' (*HighPoint*, 1986e, p. 2) from inside the organisation out. In this way, it would be second-nature to the crew to dispense the same 'courteous' dispositions while at work, replicating a polite society in the air that passengers were familiar, and could be comfortable, with.

A 1985 mini-stage production put up by SIA's cabin crew at its annual 'Group Courtesy Sketch Contest' is an apt demonstration of this form of self-disciplining that SIA was trying to inculcate among crew members. Consider this church wedding portrayal in '"I" do!... I do?':

> A couple... is about to get married and the in-laws, played by [In-Flight Supervisor] Robert See and [Flight Stewardess] Irene Rajendram, start insulting each other, almost causing the ceremony to come to a complete halt before it even commences. To add to the chaos, they have a drunkard for a best-man... and a bridesmaid Cyndi Lauper-style... The priest... with the assistance of a hunchback for an alter-boy *[sic]*... finds difficulty in conducting the matrimonial ceremony amidst all the hilarious disharmony and discourtesy in the church. Of course, it all ends happily... when the in-laws realise how being discourteous can affect any situation, even the future lives of their son and daughter.
>
> (*HighPoint*, 1985b, p. 7)

The contents of this sketch exceeded a simple message about the hilarity of 'disharmony' and 'discourtesy'. It was laden with metaphors and scenarios that might prove threatening to SIA's in-flight comfort. While at the centre was the couple/passengers wanting to commence a(erial) life together, in-laws/fellow crew who could not get along threatened to spoil the ceremony/in-flight service, despite the best intentions of the priest/pilot. Added to this were unsavoury types to be frowned upon in SIA: a ghastly-looking attendant and a hunchback, thrown together with a drunkard for chaotic effect. Ultimately, it was the in-laws' realisation that 'being discourteous can affect any situation' that the future of their children/passengers could get on. Enacted by *and for* the crew, the drama provided an object lesson about the fragility of calming comfort in SIA's cabins. When sullied by poor collegiate relations, it could spill over to manifest itself as 'discourtesy' to passengers, and make their discomforting aerial existence feel that much worse.

'Man *[sic]* is a social creature. As such, he *[sic]* responds to emotions. Any expression you make result in a reaction from the other' (*HighPoint*, 1985c, p. 10). This admonition captures the service canon that Singapore Girls were to abide by, signalling that every move they made was not only interpersonally infectious, but also needed to be acted out in accordance to the affective dimensions deemed acceptable by passengers. What SIA needed was, in short, an infusion of its cabins with something more than just captivating aesthetics, to also include a palpable sense of (anti-Oriental) calm, devoid of any disturbance or trace of unreliability that might be evoked through polluted speech, uncultured habits, or discourteous propensities. The success with which the air could be rid of such 'unfamiliar' demeanours in turn depended on the extent to which 'real-life' flight attendants could live up to the expectations of the Singapore Girl, who must not only dispense the usual good service associated with her, but also efface all 'tropical' traits apart from her 'native' charm. By performing a multi-layered form of emotional labour that was both visceral and cultural, she could then help conjure another kind of comforting ambience onboard, which could be traded for passengers' ease and satisfaction, and, above all, her country's assured air links.

Finer service

Another series of in-flight enhancements SIA made through the Singapore Girl pertained to the airline's later efforts to heighten senses of comfort for, especially, economy-class passengers, who did not have access to premium offerings such as better-designed seats, lounges and 'slumberettes'. This impetus to improve in-flight experience for the mass market was made even more pressing by the fact that 'other [airlines], particularly in Asia, also [now] boast of good inflight service' (*Outlook*, 1986a, p. 1), some of which even adopted similar Oriental motifs (e.g. Malaysia and Thai). As then-Managing Director, Cheong Choon Kong admitted, 'the Oriental culture is not exclusively Singaporean, so by itself that does not explain why our service is superior' (*Outlook*, 1986a, p. 1). A new compensatory tactic was thus needed – something a little *more than* 'Oriental', where the Singapore Girl would become even more animated and personable in-flight comforters. As another *Outlook* issue announced (1986b, p. 11), '[t]o improve passenger relations and service, Cabin Crew [Division] has adopted the theme "PR in the Cabin" for 1986 [where] Crew are encouraged to interact more with passengers and to constantly attend to their needs'. In an extension of the Singapore Girl's line of duty, succouring passengers by simply being 'courteous' to them was now no longer sufficient. What was needed was for her to energise an increasingly standard cabin atmosphere, with proactive passenger outreach and a quickened desire to relieve passengers of their unease.

This strategy first found impetus in supposedly deteriorating service standards some one decade after the airline's inception. From the outset, SIA had been an assiduous monitor of its crew's performance, and had kept a barometric record of all passenger feedback in a process akin to a Taylorist form of scientific management applied to its employees' work (Cresswell, 2006). Periodically, customers' commendations were conveyed to staff to congratulate them; but in the mid-1980s, the company had more sobering news. At the 23rd SIA Group Business Meeting, it was declared 'that our legendary inflight service appears to be losing its shine… The number of compliments received has come down significantly since December 1983, and complaints have increased' (*Outlook*, 1986c, p. 5). One explanation for this 'alarming' downward trend was the perception that Singapore Girls were slipping in a particular measure known as 'Friendly Helpful Inflight

Service' (*HighPoint*, 1986f). Negative feedbacks, which were published in every issue of *HighPoint*, starting 1988, pointed out the failings of (mostly female) crew to take the initiative to address passengers' unsaid needs or feelings, making them appear uninterested or even rude. To remedy the situation, SIA's flight attendants needed to be retrained in ways that would render them (even) more amicable and interpersonally adept, thereby allowing them to reclaim the cabin as a domain of superior comfort.

Little details mattered in this new crusade to win hearts and minds. For starters, the airline sought to re-orient the cabin as a space that was less tiresome to navigate for certain groups of needy passengers. One such group was mothers with infants, who, presumably, would find the discomforts of flying more acute. As this *Outlook* (1986d, p. 6) article entitled 'Caring for babies – the Singapore Girl way' unveils,

> Mothers travelling with their babies can count on our cabin crew for help in seeing to the needs of tiny tots. Our cabin crew, especially the flight stewardesses, are trained to pay special attention to mothers with young children.

At each pre-flight briefing, female crew watched a four-minute video of 'simulated situations' where extra attention could be bestowed on infant-totting mothers. The scope of this new service included: helping the mother carry and stow her luggage; holding the baby while the mother put on the seat belt; fixing the bassinet and helping to place the baby into it; informing the mother of the types of baby food available; frequently checking with her on whether additional help was needed; and carrying the infant while she used the lavatory. By enrolling the mother's fluctuating moods into their service scripts, the Singapore Girls were expected to juggle an even broader job scope than before, working the cabin with an anticipatory poise.

Such special attention was extended to other passengers too. For the little ones, not only would they now be pacified with sweets and toys, 'cabin crew are encouraged to cuddle the babies and play with the older children whenever time permits' (*Outlook*, 1986e, p. 6). Here, the Singapore Girl was transformed into a 'Big Sister' persona, whose comforting labours included babysitting and relieving parents travelling with children. Not to be left out, the solitary traveller was also to be made recipient of a more personalised touch. For grieving passengers (such as those travelling to a funeral), the Singapore Girl was ready to lend a listening ear as 'a counsellor and a friend' (*Outlook*, 1990, p. 7), extending a different kind of comfort not usually associated with flying. As part of its continuing 'PR in the Cabin' campaign, the company further assigned 16 'service indicators' that flight attendants could monitor themselves with, including to 'show a warm and inquiring attitude at all times'; 'be patient and sensitive to passengers in difficulty'; and 'take an active interest in the well-being of passengers' (*Outlook*, 1989, p. 11). As then-Director of Cabin Crew, Theong Tjhoen Onn explained, the way 'to progress further in our relentless pursuit of inflight service excellence' was to 'bring into sharper focus the finer points of service' (*Outlook*, 1989, p. 11). Practically for Singapore Girls, this meant that their comforting labours would now need to be sped up, to keep pace with the acceleration of competition in SIA's unfriendly skies.

To further raise productivity, SIA had additionally taken steps to ensure that these ideals were internalised by crew. Not coincidentally, a number of staff awards were instituted during this period to acknowledge and motivate flight attendants to excel even more. One of these was the Managing Director's Awards for Special Acts of Service, which was first presented to five crew members at the inaugural Cabin Crew Business Meeting in 1988. With stories ranging from a crew helping a passenger retrieve a wedding ring from the toilet, to two stewardesses nursing a feverish passenger before accompanying him to a medical centre upon arrival, the event was a showcase of how ordinary employees could have gone the extra mile to comfort passengers. As the company urged,

> [g]ood service does not end at the aircraft door: it is not limited by laid-down procedures; it extends beyond the normal service expected of our crew; it requires an exercise of initiative; it displays that extra attention; and, most of all, it is driven by genuine care for the passenger's comfort.
> (*Outlook*, 1988, p. 8)

At its crux was SIA's belief in re-engineering the aircraft cabin into an open-ended service space with the customer at the centre. Specifically, the Singapore Girl would help facilitate a service space that was finer and more efficient than any other, so that it was easier for comfort to take root.

Conclusion

While most studies on mobilities have focused on the banal experiences of transport users (Binnie et al., 2007), this chapter has adduced an example of how onboard experiences may be prefabricated too, as 'worlds' to be inhabited. As the case of SIA exemplifies, feelings of (dis)comfort in transit – as a form of such experiences – are not emotions that are simply allowed to flourish in any contingent way. Rather, they are pre-empted and shaped by transport providers to encourage, steer, and (hopefully) enable particular affects to take hold among customers. In aviation, these measures moreover do not just take the form of providing a conducive setting, or installing ergonomic seats (cf. Jain, 2011) to ease the burdens of long-haul travel, but may also implicate transport workers and their emotional labours, to help alter and enhance the socialities and moods onboard (cf. Bissell, 2010). Indeed, it was exactly through these strategies of enabling comfort that SIA was able to offer an aerial product that was competitive enough to make it a viable business. Insofar as they allowed Singapore to achieve its goal of drawing traffic to/through itself, comfort here is also arguably as much a political end-goal as a feeling.

What is more remarkable about this instance of comfort design is that it simultaneously entails particular (re)interpretations of what 'comfort' means. Beyond common associations of in-flight (dis)comforts with the cabin, its materialities and artificial environments (Bissell, 2015; Budd et al., 2011a), comfort clearly took on different shades over time in SIA's early formulations. Within the period studied, it had harboured notions of the exotic gentleness of the Singapore Girl; a calm characterised by familiarity, homeliness and graciousness; and the reassuring presence of carers whose job was to tend to passengers, pacify kids, and nurse the afflicted. All these were not pre-existing benchmarks of comfort, but were creations that had emerged in sync with the evolving tastes and expectations of travellers of that time. Not only does this demonstrate the unfixed ways (dis)comforts are encountered (Bissell, 2008), it also signals how new senses of the same can be invented, produced, and activated, as bodies are re-configured to relate with each other in novel ways. In surfing the changing tides of competition, SIA had exactly sought to induce many 'Asian' comforts through these means, pre-supposing a form of experiential difference inseparable from its cultural innovations.

Taken together, this renders the transport provider's actions that much more political, being multiply enfolded with techniques of affecting passengers and of disciplining their own workers to exact those experiences. Taken beyond the context of SIA, transport workers – including those of other vehicular types – therefore also ought to be more thoroughly investigated for how they are deployed on purpose to conjure particular ways of being on the move, in tandem with, and often preceding, passengers' practices. Neglecting these intentions and designs would not only give excessive credit to travellers for the derivation of complex mobile affects in transit. It would also allow the political methods transport providers use to socialise us all to escape vital scrutiny and critique.

Note

1 Male stewards made a late entrance in Singapore in 1961, initially as staff of Malayan Airways. In contrast to the Singapore Girl – touted the 'epitome' of SIA's cabin service – the role of the male crew was deferentially only that of 'ably complement[ing]' the former (*HighPoint*, 1994, p. 3).

References

Adey, P., Bissell, D., McCormack, D. and Merriman, P. (2012) 'Profiling the passenger: mobilities, identities, embodiments', *Cultural Geographies* 19(2): 169–193.

Adey, P. and Lin, W. (2014) 'The social and cultural geographies of air-transport', in A. Goetz and L. Budd (eds.) *Geographies of Air Transport*. Farnham: Ashgate, pp. 61–71.

Allen, R. (1990) *SIA: Take-Off to Success*. Singapore: Singapore Airlines.
Ashmore, P. (2013) 'Slowing down mobilities: passengering on an inter-war ocean liner', *Mobilities* 8(4): 595–611.
Barry, K. M. (2007) *Femininity in Flight: A History of Flight Attendants*. Durham, NC: Duke University Press.
Binnie, J., Edensor, T., Holloway, J., Millington, S. and Young, C. (2007) 'Mundane mobilities, banal travels: editorial', *Social & Cultural Geography* 8(2): 165–174.
Bissell, D. (2007) 'Animating suspension: waiting for mobilities', *Mobilities* 2(2): 277–298.
Bissell, D. (2008) 'Comfortable bodies: sedentary affects', *Environment and Planning A* 40(7): 1697–1712.
Bissell, D. (2010) 'Passenger mobilities: affective atmospheres and the sociality of public transport', *Environment and Planning. D: Society and Space* 28(2): 270–289.
Bissell, D. (2014) 'Encountering stressed bodies: slow creep transformations and tipping points of commuting mobilities', *Geoforum* 51: 191–201.
Bissell, D. (2015) 'Virtual infrastructures of habit: the changing intensities of habit through gracefulness, restlessness and clumsiness', *Cultural Geographies* 22(1): 127–146.
Bissell, D., Hynes, M. and Sharpe, S. (2012) 'Unveiling seductions beyond societies of control: affect, security, and humour in spaces of aeromobility', *Environment and Planning D: Society and Space* 30(4): 694–710.
Boris, E. (2006) 'Desirable dress: rosies, sky girls and the politics of appearance', *International Labor and Working-Class History* 69: 123–142.
Budd, L. (2011) 'On being aeromobile: airline passengers and the affective experiences of flight', *Journal of Transport Geography* 19(5): 1010–1016.
Budd, L., Bell, M. and Warren, A. (2011a) 'Taking care in the air: jet air travel and passenger health, a study of British Overseas Airways Corporation (1940–1974)', *Social History of Medicine* 25(2): 446–461.
Budd, L., Warren, A. and Bell, M. (2011b) 'Promoting passenger comfort and well-being in the air: an examination of the in-flight health advice provided by international airlines', *Journal of Air Transport Management* 17(5): 320–322.
Chan, D. (2000) 'The story of Singapore airlines and the Singapore girl', *Journal of Management Development* 19(6): 456–472.
Chong, M. (2007) 'The role of internal communication and training in infusing corporate values and delivering brand promise: Singapore Airlines' experience', *Corporate Reputation Review* 10(3): 201–212.
Chua, B. H. (1995) *Communitarian Ideology and Democracy in Singapore*. London: Routledge.
Crang, M. (2002) 'Between places: producing hubs, flows, and networks', *Environment and Planning A* 34(4): 569–574.
Cresswell, T. (2006) *On the Move: Mobility in the Modern Western World*. London: Routledge.
Gibson, T. and Harrison, M. (2005) 'Aviation medicine in the United Kingdom: early years, 1911–1919', *Aviation, Space and Environmental Medicine* 76(6): 599–600.
Gottdiener, M. (2001) *Life in the Air: Surviving the New Culture of Air Travel*. Lanham, MD: Rowman and Littlefield Publishers.
Govil, N. (2004) 'Something spatial in the air: in-flight entertainment and the topographies of modern air travel', in N. Couldry and A. McCarthy (eds.) *MediaSpace: Place, Scale and Culture in a Media Age*. London: Routledge, pp. 233–252.

Hart, B. and Bongusto, F. (1979) *You're a Great Way to Fly – Singapore Girl* [music record]. Singapore: Singapore Airlines.
Hayes, R. and Tiffney, C. (2007) *Airline Confidential: Lifting the Lid on the Airline Industry*. Stroud: Sutton Publishing.
HighPoint (1983) 'Beauty seminar for 75 stewardesses', *HighPoint: SIA Flight Crew Newsletter* 1(2): 24-24.
HighPoint (1984) 'Announcing – a fresh approach', *HighPoint: SIA Flight Crew Newsletter* 2(3): 19-19.
HighPoint (1985a) 'FSS recruitment drive', *HighPoint: SIA Flight Crew Newsletter* 3(2): 20-20.
HighPoint (1985b) 'SIA Group Courtesy Sketch Heats', *HighPoint: SIA Flight Crew Newsletter* 3(8): 7-7.
HighPoint (1985c) 'Service with courtesy in Cabin Crew Operations Department', *HighPoint: SIA Flight Crew Newsletter* 3(5): 10–11.
HighPoint (1986a) 'Beauty and grooming consultation', *HighPoint: SIA Flight Crew Newsletter* 4(3): 18-18.
HighPoint (1986b) 'Natural beauty aids', *HighPoint: SIA Flight Crew Newsletter* 4(5): 17-17.
HighPoint (1986c) 'Social etiquette courses', *HighPoint: SIA Flight Crew Newsletter* 4(1): 3-3.
HighPoint (1986d) 'Cabin crew courtesy corner', *HighPoint: SIA Flight Crew Newsletter* 4(1): 5-5.
HighPoint (1986e) 'Who is a courteous person?', *HighPoint: SIA Flight Crew Newsletter* 4(6): 2-2.
HighPoint (1986f) 'Moves to improve inflight services', *HighPoint: SIA Flight Crew Newsletter* 4(1): 15-15.
HighPoint (1990a) 'Slim is still in', *HighPoint: SIA Flight Crew Newsletter* 8(2): 14-14.
HighPoint (1990b) 'How well do you speak?', *HighPoint: SIA Flight Crew Newsletter* 8(1): 4-4.
HighPoint (1994) 'Inflight service victim of its own success?', *HighPoint: SIA Flight Crew Newsletter* 10(9): 3-3.
Hochschild, A. R. (1983) *The Managed Heart: Commercialization of Human Feeling*. Berkeley: University of California Press.
Jain, J. (2011) 'The classy coach commute', *Journal of Transport Geography* 19(5): 1017–1022.
Johnson, M. (1943) 'The history and development of aviation medicine', *Aviation Medicine* 35(6): 194–199.
Law, E. H. and Cooperrider, N. K. (1974) 'A survey of railway vehicle dynamics research', *Journal of Dynamic Systems, Measurement, and Control* 96(2): 132–146.
Lee, P. S. C. (1993) 'Blame the girls too', *The Straits Times*, 28 May, Singapore: Singapore Press Holdings, p. 2.
Lim, C. B. (2000) Oral History Interview – 05 September [Transcribed by K. Koh in October 2009], Record held at the Singapore National Archives, Accession Number 002358/03.
Martin, C. (2011) 'Desperate passage: violent mobilities and the politics of discomfort', *Journal of Transport Geography* 19(5): 1046–1052.
McCormack, D. P. (2008) 'Engineering affective atmospheres on the moving geographies of the 1897 Andrée expedition', *Cultural Geographies* 15(4): 413–430.

Murphy, A. (1998) 'Hidden transcripts of flight attendant resistance', *Management Communication Quarterly* 11(4): 499–535.
Neveu, A. J., Koppelman, F. S. and Stopher, P. R. (1979) 'Perceptions of comfort, convenience, and reliability for the work trip', *Transportation Research Record* 723: 59–63.
Office of the Commissioner-General (1962) 'Visit to Kuala Lumpur', Office of H.M. commissioner-general for U.K. in S.E.A., 17 January, Record held at the National Archives, Kew, Serial Number BT 245/1001.
Omelia, J. and Waldock, M. (2006) *Come Fly with Us! A Global History of the Airline Hostess*. Portland, OR: Collectors Press.
Outlook (1980) 'Setting high standards for cabin crew training'. *Outlook: House Newspaper of the SIA Group*, July Issue MC(P)31/1/80: 8–9.
Outlook (1986a) 'Superior training keeps us ahead', *Outlook: House Newspaper of the SIA Group*, September Issue MC(P) 255/1/86: 1–5.
Outlook (1986b) 'Improving our product and services', *Outlook: House Newspaper of the SIA Group*, August Issue MC(P) 255/1/86: 11-11.
Outlook (1986c) 'A public company – what does it mean?', *Outlook: House Newspaper of the SIA Group*, March Issue MC(P) 16/3/85: 5-5.
Outlook (1986d) 'Caring for babies – the Singapore girl way', *Outlook: House Newspaper of the SIA Group*, June Issue MC(P) 255/1/86: 6-6.
Outlook (1986e) 'No child's play – catering to kids onboard', *Outlook: House Newspaper of the SIA Group*, June Issue MC(P) 255/1/86: 6-6.
Outlook (1988) 'Rising beyond the call of duty', *Outlook: House Newspaper of the SIA Group*, September Issue MCI(P) 43/1/88: 8–9.
Outlook (1989) 'Cabin Crew takes a closer look at service', *Outlook: House Newspaper of the SIA Group*, February Issue MCI(P) 124/1/89: 10–11.
Outlook (1990) 'SBC spotlights the Singapore Girl', *Outlook: House Newspaper of the SIA Group*, September Issue MCI(P) 58/90: 7-7.
Raguraman, K. (1997) 'Airlines as instruments for nation building and national identity: case study of Malaysia and Singapore', *Journal of Transport Geography* 5(4): 239–256.
Said, E. W. (1978) *Orientalism*. London: Routledge and K. Paul.
Shaw, H. and Hesse, M. (2010) 'Transport geography and the "new" mobilities', *Transactions of the Institute of British Geographers* 35(3): 305–312.
Simonsen, D. G. (2010) 'Transitrum. Flykabiner og supermodernitetens ikke-steder', *Scandia: Tidskrift för Historisk Forskning* 74(2): 103–126.
Tan, S. (2000) 'Your choice: vote your moodle – Singapore Girl', *The Straits Times*, 9 January, Singapore: Singapore Press Holdings, p. 4.
Teo, T. S., Srivastava, S. C. and Ho, C. K. (2006) 'The trident model for customer-centric enterprise systems at comfort transportation, Singapore', *MIS Quarterly Executive* 5(3): 109–124.
Tham, K. W. (1983) 'National campaigns in Singapore politics', Unpublished Thesis, Department of Political Science, National University of Singapore. Singapore: National University of Singapore.
The Business Times (2007) 'To change or not to change: does the Singapore Girl need a makeover?', *The Business Times*, 20 February, p. 4.
The Sunday Mail (1986) 'Chopsticks banned at Mrs Nicholas' school for Singapore', *Sunday Mail (Queensland)*, 2 November, no page.

Watts, L. (2008) 'The art and craft of train travel', *Social and Cultural Geography* 9(6): 711–726.
Wee, L. (1999) 'Flying high with S'pore Girl', *The Straits Times*, 13 October, p. 3.
Whitelegg, D. (2007) *Working the Skies: The Fast-Paced, Disorienting World of the Flight Attendant. New York: New York University Press.*
Wilson, H. F. (2011) 'Passing propinquities in the multicultural city: the everyday encounters of bus passengering', *Environment and Planning A* 43(3): 634–649.
Wirtz, J. and Johnston, R. (2003) 'Singapore Airlines: what it takes to sustain service excellence–a senior management perspective', *Managing Service Quality* 13: 10–19.

Section three

Materiality and texture

10 Comfort, identity and fashion in the post-socialist city

Mark Jayne

Introduction

This chapter shows that study of geographies of comfort, consumption and consumer culture focused on materialities, assemblages and context has much to offer theoretical and empirical understanding of the ways people claim a 'right to the city'.[1] The concept developed by Lefebvre (1968) in the late 1960s has been recently rediscovered by critical urban theorists and has also become a rallying call for transformative political and social movements (Harvey 2008; Brenner et al. 2012). Schmid (2012: 58), for example, suggests that the legacy of Lefebvre's writing on collective action has been to infuse critical urban thinking with an interest in 'everyday life, the banal, the ordinary ... [and that] changing everyday life: this is the real revolution ... [where] any point has the potential to become central and be transformed into a place of encounter, difference and innovation'. At the same time, the emergence of writing on materialities and assemblages relating to 'collective consumption' such as housing, water, sanitation and so on has taken a different route towards similar critical ground. The post-structuralist writing of theorists inspired by Latour (2005) and Deleuze and Guattari (1988) has highlighted how human and non-human actors come together in urban political and social struggles (McFarlane 2009; Farias and Bender 2010, 2011; Dovey 2011). In this chapter, we work at the intersection of this writing and in doing so we also engage with Lefebvre's (2009: 148) argument that 'limiting the world of commodities is key to the advancement of radical democracy' (Lefebvre 2009: 148). Focusing on 'individualized' clothing consumption we investigate productive 'collective' social relations and experiences through the wearing of 'comfortable clothing'.

More specifically we focus on the relationship between fashion and identity in public spaces in Petržalka, a high-rise housing estate in Bratislava, Slovakia.[2] We show that wearing 'comfortable' clothing by diverse socio-economic groups and across generations in Petržalka is a way that citizens make public spaces more 'homely', an expression of collective local identity and 'belonging', formulated in opposition to individualized 'petit-bourgeois' consumption dominant elsewhere in the city. The chapter begins by reviewing theoretical and empirical writing from across the social sciences relating to fashion and identity. We then argue that a focus on comfort, fashion and urban identity through the lens of assemblages, materiality and context offers more complex political, economic, social, cultural and spatial analysis. To that end, subsequent sections engage with geographies of wearing comfy clothing during both the socialist era and the emergence of variegated capitalism since 1989. We conclude by discussing how the wearing of comfortable clothing is an everyday 'political' response to changing temporal and spatial imaginaries and experiences of political and economic change as citizens claim a material 'right to the city'.

Theorising geographies of comfort, urban fashion, consumption and consumer culture

It will be of no surprise to fashion, consumption or urban theorists that wearing 'comfortable' clothing in public spaces; pyjamas, sportswear, and other clothing most often thought as only being acceptable to wear 'at home', is a feature of urban life around the world. For example, in Shanghai, Iossifova (2011: 202–203) describes how local customs such as 'wearing pyjamas in public, drying blankets on the street and spitting, for instance [are] being portrayed as backward or rural … [edit] ascribed to the residents of 'old residential areas, by the Government and the media' (see Figure 10.1). In the UK there have been numerous examples of 'comfy clothing' causing controversy. For example, in Liverpool it is popular for women, in preparing for the weekly 'big night out' on Fridays and Saturdays, to wear comfortable clothes *and* walk around the city with 'curlers in their hair' (see Figure 10.2). In Middlesbrough (BBC 2011) parents who wear pyjamas when taking children to school and who attend school parent/teacher meetings in their nightwear have been criticised, with it being noted that they 'drop them [their children] off in the morning and are collecting them wearing the same pyjamas' suggesting that this practice is creating a 'bad impression'. In Belfast, jobseekers have been banned from wearing pyjamas at the local social security office, and in Cardiff a supermarket has outlawed 'nightware' in its aisles (*Irish Central* 2012). The trend of wearing comfy clothing in public spaces is also prevalent in the affluent upper east side of New York City with fashion-conscious women and teenagers parading their Louis Vuitton and other designer pyjamas on the streets of Manhattan. This fashion is not however new, with a headline in the *New York Times* in 1929 highlighting, 'Court sanctions pyjamas in the street'. The article discusses how a man was arrested only to be released by a judge who warned the police that 'Neither you nor I are censors of modern fashion' (Manjoo 2012). While negative moral values assigned to these consumption practices are generally articulated around the lines of class, gender and ethnicity the geographies relating to wearing comfy 'clothes' in urban public spaces clearly resonate around the world and requires further consideration.

Figure 10.1 A scene from the past? Wearing pyjamas in public spaces in Shanghai.

Figure 10.2 Scouse women 'at home' in Liverpool City Centre.

Writers such as Veblen (1898), Simmel (1957), Benjamin (1982), Hedidge (1979), Wilson (1985) and Maffesoli (1994), to name but a few, have long articulated the spatial, social and psychological processes relating to bodily adornment. Fashion has been described as 'decadent and frivolous' but nonetheless part of a 'civilization' process, that Simmel, describes as 'dialectic performance of fashion', a continual search for 'newness' (1957: 106). More recently, theorists have debated 'the end of individualism' and the emergence of a 'time of tribes' where membership is not necessarily orientated around traditional social structures but consumption, fashion and lifestyle (Maffessoli 1996). This work has signposted the relationship between fashion and specific urban spatial, cultural and historical contexts and the importance of pursuing 'comparison, emulation and differentiation [that] are most noticeably apparent in the rapid changes that characterize systems of industrial production' (Cannon 1998: 23).

Elizabeth Wilson (1985), for example, considers fashion as essential to modernity, spectacle and mass communication. A key focus of Wilson's (1985) work has been to identify how fashion emerged in archetypal modern cities. In nineteenth-century Paris and twentieth-century New York, Wilson describes how industrialism brought about new complicated bourgeois 'codes of dress', available to buy in arcades and department stores, and later in decentred suburbia, where 'exclusivity and chic belonged to metropolitan life; dowdiness to the provincial backwaters' (Wilson 1985: 154). Wilson maps out histories of fashion from the pre-industrial world, through the excesses of aristocrats, and the diverse range of trends associated with the ascendancy of urban middle-classes. In terms of the latter, Wilson describes the emergence of a range of materials (including cotton, calico and muslin) related to ideals of thrift, work and sobriety and particularly the wearing of black suits by men as well as the work of designers such as Coco Channel and Claire McCardell, who from the 1930s onwards revolutionized the way that women dressed through the use of design and fabric. In particular, Wilson (1985: 41) highlights McCardell's use of 'tights, flat shoes and soft easy styles', as being key points in the emergence of more 'comfortable' ways of dressing.

Recent writing has built on these ideas in order to consider the relationship between fashion production and consumption in specific spaces and places (see for example, Entwistle 2000; Breward et al. 2006). Studies have focused on a diverse array of contexts and topics, for example, fashion in junior schools in UK (Swain 2002), fashion in late Qing period in Shanghai with reference to sexuality, desire and mixing of 'western' and Chinese styles (Zamperini 2003) and 'postmodern' style in Milan (Bovone 2006). Bernard (1996: 65–66) considers the pleasures of feeling clothes on the skin and Colls' (2004) discusses the ways women feel comfortable in their clothes and in store changing rooms. However, while there is a diverse range of writing about fashion and identity located in specific urban spaces and places there has been a relative lack of engagement with the symbolic, material, emotional and embodied relationships between fashion *and* the city. A notable exception is the work of Borden (2001) who focuses on skateboarders' baggy jeans and hooded tops allowing ease of movement facilitating a material relationship between their bodies, skateboarders and urban infrastructure. Holliday (1999), Moran and Skeggs (2004) also discuss the consumption of comfortable clothes in domestic spaces for lesbians and gay urbanities, who not always able to visibly express their sexuality through fashion in public spaces (see Shove 2003; Miller 2008). Such work compliments Mort's (1996) depiction of hidden (and not so hidden) codes of fashion that constitute masculinity and sexuality in urban spaces and places.

It is against the backdrop of this writing on geographies of fashion and consumer culture that we respond to calls to re-materialize urban studies. Latham and McCormack (2004: 703) suggest that a focus on assemblages of human and non-human actors can 'multiply the pathways along which the complex materialities of the urban might be apprehended'. Drawing on the post-structuralist writing of Latour (2005) and Deleuze and Guattari (1988), work on urban networks, practices and spaces have sought to speak 'not to static arrangement or a set of parts, whether organized under some logic or collected randomly, but to processes of arranging, organizing, fitting together ... where assemblage is a whole of some sort that expresses some identity and claims to a territory' (McCann and Ward 2011: 12). Engagement with this writing has been argued to have dramatically 'changed urban research' (Farias and Bender 2010), allowing focus on interaction of human and non-human actors in order to better understand the spatial complexity and openness of cities and the ways in which cities are assembled and dis-assembled. Theorists have thus started to engage with 'the multiple spatial networks that any city is embroiled in, and to ... [allowing consideration of] the full force of those networks and their juxtaposition in a given city upon local dynamics' (Amin 2002: 112).

For example, Colin McFarlane (2009) considers composites of place-based exchanges of ideas, knowledge's, practices, materials and resources. McFarlane (2009) describes 'collective consumption' such as housing and sanitation in terms of re-assembled socio-material practices that are diffuse, tangled and contingent being constituted by a diverse number of groups, collectives and by extension agencies in order to conceive of 'power as multiple co-existences ... [and that] assemblages denote not a central governing power, nor a power equally distributed, but power as plurality in transformation' (2009: 562). McFarlane (2011) focuses on group exchanges involving people, materials, resources, histories and struggles, and calls for an approach where multiple concerns of space and power are understood as open to multiple spatial imaginaries and practices.

Theoretical and empirical progress in studying materialities and assemblages has not however been without controversy. For example, in calling for assemblages research to take seriously underlying logics and inequalities of capitalist accumulation, Brenner et al. (2011: 227) welcome the 'innovative, intellectually adventurous impulse behind recent assemblage-theoretical interventions'; is cautiously optimistic about the empirical foci, but suggests that assemblage writing downplays the 'context of contexts' and thus fails to adequately grasp how capitalism shapes contemporary urbanization. Other writers have offered varying levels of support for assemblage thinking and its critics. For example, Dovey (2011) applauds the possibilities that assemblages has to 'nudge' critical urban theory outside a political economy framework, asserting that understanding of urbanization is more complex than focusing on capitalism as a root cause of all urban practices and processes. Dovey further argues that assemblage urbanism seeks to overcome tendencies in critical urbanism to resort to hierarchies of scale that valorize the large (e.g. global capital) over the small in a manner that grasps the complexities and messiness of 'the urban'. Simone (2011a: 330) similarly argues that capitalist logics do not provide exhaustive accounts, and that writers must consider how urban life in specific sites takes place and 'gets done', 'through a lens of domination, commodification and dispossession … [via concern with] iterative and opaque processes of adaptation, hesitation and collaborations'.

In response to Brenner et al. (2011), McFarlane (2011) suggests that study of materialities and assemblages does not sidestep or displace geographies of political economy, pointing to 'artificial divisions' in Brenner et al.'s critique of a divide between political-economy and post-structural thinking that they impose in their review of assemblages writing. McFarlane further points to the ways in which assemblages offer a set of approaches to challenge capitalism by understanding the effect of socio-material practices rather than as an underlying or essential logic, in short engaging with the 'context of contexts' without erasing the complexity and contingency of urban change and struggle. To a large degree, Brenner et al. are overly quick to criticize a body of writing that is at a relatively earlier stage of development in comparison with the traditions of 'geopolitical economy' research that they speak from. Moreover, Brenner et al. (2011) sidestep the work of theorists such Rao et al. (2007); De Boeck (2011) and Simone (2011a) who have been interested in urban assemblages as a vital component in understanding spatial tactics of capitalist accumulation. These theorists focus on the coming together of architectural materials and the re-assembling of urban imaginaries through practices of construction and occupation, production and consumption, forms of sociabilities, social relations, and politics in terms of specific spaces and places and their broader 'contexts'. For example, Simone (2011a: 364) highlights how 'ports, municipal administrations, bus terminals, 'offshore' industrial plants, back office processing zones, large-scale low-income and middle class housing developments and universities … [are] domains where politics, culture, economy and techniques are potentially folded in many different ways and as sites of possibility to take urbanism in different directions'. In the remainder of the chapter we contribute to this theoretical and empirical terrain.

Geographies of comfort, identity and fashion: public/private space in socialist Petržalka

In unpacking interpenetrations of political, economic, social, cultural and spatial practices and processes bound up with wearing comfortable clothing in Petržalka we also contribute to the advancement of social and cultural understanding of post-socialism. Research into post-socialism and post-socialist urbanity has, especially since the 1990s, predominately focused on theorizing political and economic changes associated with the emergence of capitalism. Although offering much to understanding urban change in Central and Eastern Europe and other post-socialist countries since the fall of the Berlin Wall, Ferenčuhová (2011: 66) argues that local thinkers have lacked 'ambition to contribute to urban theoretical debates' more broadly. As Ferenčuhová and Jayne (2014) suggest theoretical and methodological advances through critical Marxism, feminism, post-colonial and post-structural approaches have had little impact on the direction of (post)socialist urban thinking. Such comments notwithstanding, a small body of writing has sought to address this imbalance. For example, researchers have analysed changing urban landscapes, focusing on (re)definitions of social and cultural forms and practices in relation to geographies of political and economic change (Bitusikova 1998; Stenning 2000; Smith 2002; Czepczynski 2008; Ferenčuhová 2009, 2011). Horschelmann and Stenning (2008: 355) have also championed ethnography in post-socialist studies in order 'to answer some of the difficult demands of researching the trans-local flows and connections in which most lives are enmeshed globally, but can none the less *locate* itself along particular intersections, thus maintaining awareness of positionality and not eroding the difference that space makes'. When read together this work shows how complex spatialities of post-socialism can be productively considered through 'lived experience' (Horschelmann and Stenning 2008: 345).

In these terms, while Petržalka has attracted academic interest as an archetypal Central Eastern European Socialist era mass housing estate built during the 1970s (Stenning et al. 2010), it is important to view Petržalka as a symbolic site of nation building following the unification of Slovakia and the Czech Republic after World War Two. During this period, urban planning for the newly unified nation of Czechoslovakia was focused the capital city of Prague followed by attention to Brno, the second largest Czech city and also Bratislava. While development of Bratislava and Brno was underpinned by physical and economic growth, Ferenčuhová (2011) argues that the political imperatives for Bratislava to become the 'second city' of Czechoslovakia were vital to ensure the involvement of Slovaks in buying into a unified national identity. As part of that project Petržalka became home to 'immigrants' from all over Slovakia, relocated to Bratislava as part of state urbanisation and industrialization planning programmes. Young people and families from diverse socio-economic backgrounds populated the newly built high-rise apartments of Petržalka and while the current population of 116, 993 inhabitants can be characterised by increasing numbers of residents who are economically less well off, the demographic structure of Petržalka still reflects the socio-economic mix of residents that characterised the early years of the housing estate (SNS 2010).

Our research shows that the wearing of comfortable clothes in public spaces in Petržalka was an attempt by newly arriving residents to make public spaces 'work for themselves'. Informality and appreciation of personal 'comfort' was noted by respondents are being related to physical separation from the rest of the city (by the river Danube and a motorway) and because of the 'estates lacks of architectural beauty' (see Figures 10.3 and 10.4). This point was made by Babeta (Female, 60, Professor) who then went on to describe her own surprise while moving into an apartment in Petržalka witnessed a conversation between two neighbours dressed in negligees 'shouting' across the street, from the window of one high-rise building to another, and František (Male, 65, retired) who recounts how residents claimed the newly built quarter 'as their own':

> people present themselves as 'being in Petržalka' and they behave accordingly [edit] ... They have to try and feel at home in Petržalka and they often behave as if in their own kitchen. [edit] ... because everyone knows each other and lives so close together in the flats that all look the same ... [edit] they have no scruples to go out in tracksuits, or in shorts, to the street, or to the shop [edit] ... Petržalka, its public spaces, are part of their home [edit] ... If they went downtown [to the city centre], they wouldn't dress like that, but here, they do [edit] ... Here whether it is school kids, middle-class, or loafers it doesn't matter [edit] ... They think this is their place and they set the rules...

Figure 10.3 The motorway that connects/divides Petržalka with/from the rest of Bratislava.

Source: Mark Jayne.

Figure 10.4 The River Danube, a boundary between Petržalka and the rest of Bratislava.
Source: Slavomíra Ferenčuhová.

This quote highlights an expression of comfort as a key part of a 'collective' identity in Petržalka and the 'fuzziness' of the public and private space. The wearing of comfortable clothes thus represents and is performative of a boundary crossing, a 'moral' logic of encroachment, rather than being 'taking possession of public space in a selfish way, it's more like extending that you give your own space to the public space' (Bromley 2004: 294). Public and private thus do not emerge as neatly exclusive or exhaustive categories but where 'popular meanings can be produced through dialogical encounters... [with people looking to] the material form of the site, and its location, in order to discern the intent of the space and thus shape a moral and aesthetic response to it' (Bromley 2004: 294).

Reflecting on geographies of everyday life in the socialist era, respondents also talked about the ways in which clothes were too expensive, less fashionable that western clothes, that there was a lack of choice and moreover that clothes were often 'uncomfortable' to wear:

> in the 1960s and 1970s [edit]... When I bought Italian shoes for 400 crowns, unfortunately I couldn't afford to buy trousers at the same price, jacket at the same price, shirt at the same price, tie at the same price, raincoat at the same price so I only could buy some of these. Very few people at this time could afford to buy a whole outfit [edit] ... Then, more things became available in seventies and early eighties but were even more expensive ...[edit] to afford fashionable clothing meant not being able to afford something else. If you had a good salary you could

> afford eating and some clothes, or, if poorly paid, some people bought clothing only, and didn't eat …
>
> František (Male, 65, retired)

> women did not have boutiques here, but we always had fashion magazines. I used to make clothes for my kids and for myself [edit]… when my sister went to England [in the 1960s] they said – yes, Slovak girls are very beautiful, but they wear very ugly skirts.
>
> Babeta (Female, 60, Professor)

Babeta also commented on the texture of clothing fabrics suggesting 'the dress used to bite', and that being able to dress in 'comfy' clothing when back 'at home' in Petržalka was 'something of a relief' from having to wear 'biting' clothing when at work. In these terms, the wearing of comfortable clothing in Petržalka can be understood not only as a 'local' response to the assemblages that constituted the symbolic, built infrastructure of nationalistic Socialism and materialities of everyday life in a modernist housing estate, but to what McFarlane (2011: 219) calls cosmopolitanism as a kind of 'worldliness' which takes four relational forms:

> as 'a *knowledge*, of how difference might be negotiated or how mutuality across differences might operate; as *a disposition*, either as progressive orientation to urban cultural diversity or as regressive exclusionary sensibility replied in relation to other cultures; a *resource* as means of coping and getting by, surviving and managing uncertainty in the city; and finally as, *an ideal,* openness to and celebration of urban diversity and togetherness to be worked towards

These relational forms are clearly present in imaginaries and experience of living in Petržalka, acknowledged by both locals and non-residents and performed through materialities of 'comfortable' clothing.

For instance, respondents suggested that elsewhere in the city, such informal dressing in public spaces was frowned on as a signifier of 'lack of culture' or 'rural-like character' but in Petržalka familiarity and informality expressed via 'comfy' clothing was celebrated as a marker of identity:

> Well, yes! That relaxed attitude. Everyone knows each other. I find it fantastic [edit] … if one guy is on drugs and the other has a university degree and a good job they still know each other. They might have different social backgrounds but people have something in common [edit] … everyone is so close to each other … [edit] living together in the flats everyone lives on top of each other amongst the concrete.
>
> Lukáš (Male, 28, project manager)

Respondents also pointed to the specific human and non-human relations of physical isolation from the rest of the city, socio-economic mixing, the 'newness' and concentration of the blocks of flats, as being different to other parts of the city:

> Of course, when you are at the estate and you go to the grove or by the lake, you wear sporty clothes [edit] ... You don't go in high-heels [edit] ... from the perspective of the inhabitants, with everyone living near each other in flats, meeting on the stairs and all the concrete that surrounds, you don't need to dress up and wear those high-heels [edit] ...You wear the two-piece, or skirt when you go elsewhere, because it is part of the game in the city ...
>
> Anna (Female, 50, secretary)

> Here in Petržalka, no one takes notice of me, of course, why should they [edit] ... look at the buildings that surround you, it is not a fashionable place ... [elsewhere in the city] they pay attention to what they are wearing. [edit] ... Here, when you find someone dressed up decently, you would think he is going downtown
>
> František (Male, 65, retired)

The 'game' as noted by Anna and alluded to by other respondents acknowledges how local residents individually and collectively rejected bourgeois 'distinction' practices that have played out in urban public spaces since the late eighteenth century, discussed in detail by writers such as Veblen (1898), Simmel (1957), Benjamin (1982), Hebdige (1979), Wilson (1985) and so on. All of our respondents noted how consumption cultures and practices of 'dressing to impress' were not part of everyday life in Petržalka. For example, as Lukáš commented when discussing clothing and fashion elsewhere in Bratislava, he contrasts *Korzo* (promenading in the city centre) with the 'comfort' of Petržalka defined as 'opposite of that perfect look'.

These findings offer interesting parallels and contrasts with the work of Holliday (1999) and Moran and Skeggs (2004) who describe how, at home, lesbian and gay urbanites perform a 'politics of comfort' by using clothing to express a leisure-time based identity at home. In contrast to masking sexuality completely, or expressing sexuality through hidden codes due to pressures of being 'out' at work or in particular spaces and places in the city, Holliday (1999: 481) shows how 'at home' in a queer context 'comfort might be read as embodying *resistance* to hegemonic discourses of 'proper' behaviour and attire'. The similarities and differences of such critique with our research can be further understood by engaging with the notion 'comfort as detachment', a separation from others 'which implies a lack of necessity to worry about the world or one's position within it ... comfort as an easy unthinking state' (Holliday 1999: 490). Also reflecting on the relationship between public and private performance of identity through fashion, Moran and Skeggs (2004) describe comfort in terms of inclusion and exclusion; as defining features of identity and community, of commonality and belonging. In the first instance they argue that comfort is associated with the domestic, but it can also be linked to public and semi-public spaces such as the street, a pub, or a friend's house (2004: 83–84). Moran and Skeggs's (2004: 84) argument that 'the private and the public significance of comfort make it appear to be ambiguous and contradictory'. In these terms, the wearing of comfortable clothing by residents in Petržalka can be seen as a 'political' response to re-location to this new housing estate as part of a strategy of state socialist planning and national identity formation. Moreover, the material fabric of this new late-modern form of living, of everyday life in a de-centred socially mixed 'concrete' high-density housing estate, which markedly contrasted to dominant bourgeois city centre consumption cultures based around social distinction strategies, generated individual and collective expressions and performances of comfort and belonging.

If we develop this point with reference to McFarlane's (2011: 209) interest in 'the intensity and excessiveness of the moment', it is possible to argue that wearing of comfortable clothing in Petržalka represent a 'disruption of pattern … [which] generate new encounters with people and objects, and invents new connections and ways of inhabiting everyday urban life … [and in doing so represents] the potential of urban histories and everyday life to be imagined and put to work differently'. Particularly useful in elaborating this argument is McFarlane's (2011, citing Hardt and Negri 2009: 124) use of 'commons', as a process of becoming, a doing that constitutes 'an assemblage of affects or ways of being'. In these terms, the wearing of 'comfortable' clothing in Petržalka across diverse social groups and generations can be understood as 'a kind of gathering or multiplicities through the political work of assembly … an experimentation with cooperative spaces, processes and possibilities across multiple differences, and emerges both in relation to and in excess of assemblages of enclosure' (McFarlane 2011: 212).

However, it is important to draw attention to the agencies of such material geographies of comfort in more detail. McFarlane (2011: 221), for example describes how materiality can 'shape inequality and the prospects for resistance and alterity [and how a focus on] assemblage asks us to consider how critical praxis emerges through socio-material interaction rather than through a separation of the social and the material'. For example, the relationship between geography, fashion, identity and urban life usefully described by Sennett (2003: 39) in terms of an interchange between people and things and 'conflicting influences of individuality and conformity, change and continuity, past and future' relates to comfort, materiality and public/private in a number of different ways'. To put it another way the relationship between comfort, fashion in Petržalka can be theorized as a 'political response' that is an:

> approximation of cosmopolitanism, the assemblage imaginary recalls the concern that the 'rights to the city' but does so through a politics of recognition that has the potential implication of generating new urban knowledges, collectivities and ontologies. Assemblage's

> imaginary of gathering and composition is one vehicle through which the rights to the city might potentially be realized, whereby assemblage extends the rights to the city as a process of antagonistic composition.
>
> (McFarlane 2011: 221)

Seeking to understand the relationship between geographies of comfort, fashion and identity thus demands what Farias (2011: 365) considers as a challenge to 'notions of power as a resource of the ruling classes and of knowledge as an ideological construct that needs to be unveiled – such a thought runs the risk of silencing the heterogeneity of human and non-human actors involved in the object of critique'. As our case study research and other writing on comfort and fashion show there is a need to investigate both the nature of, and limits to, a 'political' project bound up with assemblages, materiality and public/private space as imagined and experienced in relation to 'comfort' as 'a right of access to participation'' (Amin and Thrift 2002). In the remainder of this chapter we thus question the extent to which the wearing comfy clothes in Petržalka can be theorized with reference to democratization of (non)human interactions that emerged in relation to changing temporal and spatial 'context'.

Changing context, geographies of consumer capitalism, changing clothes? Petržalka since 1989

Following the revolution in 1989, it is important to note that housing estates such as Petržalka came under a spotlight in new ways, no longer celebrated as ideological and infrastructural success stories of state socialism, but rather through critical depictions of the lives of people forced to live in 'rabbit pens' and defined as places of chaos, disharmony and discomfort (Václav Havel citied in Czepczynski 2008: 98). Unemployment, increasing poverty and criminal gangs in Petržalka and elsewhere in Bratislava ensured that for a short while after 1989, for some residents the wearing of comfy clothes became a way to avoid social distinction, rather than as a marker of belonging. More broadly, the extremes of wealth and poverty that emerged as capitalist accumulation took hold were made visible through the growth of spectacular buildings in the city centre, an influx of tourists to Bratislava's historical city centre and the rise of gated communities and affluent suburbs. Such spaces and places contrasted to Petržalka, and other quarters, now increasingly blighted by poor maintenance of buildings (due to withdrawal of state funding of infrastructural maintenance) and 'un kept' and decaying public spaces (see Figure 10.5). While everyday life and concrete materialities of Petržalka had been long derided by its residents, the growing characterization of the estate as representative of the worst kind of state-socialist urban planning from 'outside' challenged and problematized local pride in comfort and informality.

Figure 10.5 Public space in Petržalka.
Source: Slavomíra Ferenčuhová.

However, it is in these changing structural context that the continued importance of comfortable clothes for individual and collective identities in Petržalka since 1989 can be understood with reference not only to historic associations of 'homeliness' and 'belonging', but also to changes in the geographies of consumption landscape beyond the quarter. As the following quotes show, the discursive construction of spatial isolation and socio-economic mixing in Petržalka was re-imagined with reference to new consumption spaces that had appeared elsewhere in the city. Lukáš, for example, talked about the proliferation of consumer culture with reference to city centre shops and suburban shopping malls, which represented growing presence of international chains and global brands and increased 'choice' in clothing available. In a similar vein, Anna pointed to the increases in the opportunities to promenade in the city and an intensification of social differences expressed through fashion elsewhere in the city:

> After 1989, *Korzo* became even more popular [edit] ... more and more people started to go and get dressed up, and not just on the weekends [edit] ... now you see so many events in the city centre ... Christmas markets, concerts, open air cinema, New Year's Eve, or so... [edit] but now it is much easier to see differences in social background ...
>
> Anna (Female, 50, secretary)

> I left Bratislava when I was 18 [edit] ... When I came back I felt that Bratislava was dressing very chic and that people go out dressed up,

> girls wear make-up, men dress in fashionable shirts and suits even to go out to a pub, where it is not really necessary to dress that well [edit] ...
>
> (Lukáš, 28, project manager)

Responding to a question as to whether Petržalka had changed in similar ways, Lukáš suggested that:

> [Petržalka is] everything behind the Danube, everything south of the Danube. There is a clear division, us and them, and we are separated by the river. Petržalka has always been different [edit] ... We always understood it as different and they always understood it so, too [edit] ... we are still known for wearing comfortable clothing.

Such responses show that while increasing social and spatial differentiation expressed through fashion had proliferated throughout Bratislava, collective identity across socio-economic groups and generations in the wearing of comfortable clothing nonetheless remained an important marker of collective identity in Petržalka (see Figure 10.6). As such, while Maffesoli (1994) argues that 'postmodern' consumer culture has led to the emergence of fashion 'tribes', that people are free to join regardless of social-economic background, in Petržalka collective identity of 'comfy clothing' initially promoted as a response to economic and material conditions of socialism was now re-articulated because of increased social divisions associated with geographies of the emergence of consumer capitalism.

Figure 10.6 Contemporary wearing of 'comfy' clothes in Petržalka.
Source: Slavomíra Ferenčuhová.

Ash Amin's (2008: 5) post-humanist account of urban public space is useful in understanding the changing context of comfortable clothing in Petržalka, and the ways in which 'human dynamics in public space are centrally influenced by the entanglement and circulation of human and non-human bodies and matter in general, productive material culture that forms a kind of pre-cognitive template for civic and political behaviour'. Drawing on the notion of 'situated surplus' which is 'manifest in varying dimensions of compliance, as the force that produces a distinctive sense of urban collective culture and civic affirmation in urban life', Amin (2008: 8) challenges us to re-think the ways 'in which public spaces are produced and experienced through the situated multiplicity and social practice that encapsulate the rhythms of daily life in urban public spaces, by considering the resonance of collective repetition and endurance'. In these terms the blurring of public/private spaces in Petržalka can be understood by the resonance of 'situated surplus, formed out of entanglements of bodies in motion and the environmental conditions and physical experiences of tacit and neurological and sensory knowing quietly contributing to a civic culture of ease in the face of urban diversity and the surprises of multiplicity' (Amin 2008: 11). Such observations in the context of Petržalka highlight how:

> the iconography of public space, from the quality of spatial design and architectural expression to the displays of consumption and advertising along with the routines of usage and public gatherings, can be read as powerful symbolic and sensory code of public culture. It is an active code, both summarizing cultural trends as well as shaping public opinion and expectation, but essentially in the background as a king of atmospheric influence.
>
> (Amin 2008: 13)

Moreover, in order to understand the changing imaginaries and experiences of wearing comfortable clothing in the public spaces in Petržalka in terms of structural shift there is a need to respond to Brenner's (2009: 199) call to address the 'myths, reifications and antimonies that pervade bourgeois forms of knowledge … [and that] outcomes on the ground are a matter of context, shaped by material dynamics and historical legacies of individual public spaces'. As such consumption of 'comfy' clothing in Petržalka can be understood with reference to pre and (post)socialism in terms of the interpenetration of political, economic, social, cultural and spatial re-scaling(s). These include, for example, the building of a housing estate to celebrate Czechoslovakian national socialism and the response of its new residents to defining belonging and local identity; political, economic and social and cultural restructuring associated with post-socialism and outside attacks on everyday life in Petržalka, as the new Slovak nation emerged into consumer capitalism, European cosmopolitanism and proliferation of global brands and so on. In this 'context of contexts' considering the human and non-human relationships between changing discursive and differential constructions of 'comfort' and ideologies of public/private space, our research allows understanding of how a socially mixed housing estate was a site of 'civic becoming' under both socialism and consumer capitalism (Amin 2008: 22).

However, reflecting on the political project of public space, Amin (2008: 8) describes 'situated surplus', such as comfort and informality, as being 'politically modest (as sparks of civic and political citizenship) but still full of collective promise [and Amin] locates this promise, however, in the entanglement between people and the material and visual culture of public space, rather than solely in the quality of social interaction with strangers' (2008: 8). To put it another way, in suggesting that an emphasis on 'democratic politics ... actual urban situations [which] define the space of intervention for an urban democratic public, not capitalism at large' (Farias 2011: 372) allows us to understand the wearing of comfortable clothing in Petržalka as part of a 'symbolic projection', aligned with a 'strong sensory, affective and neurological response' (Amin 2008: 15). This could be criticised as being 'solidarity in a minor key' (Amin 2008: 15–17) and based on consumer culture which the quote at the beginning of this chapter by Lefebvre suggests, as working against 'radical democracy'. However, while the wearing of comfy clothes in Petržalka may indeed be 'politically modest', our research nonetheless highlights 'political' practice, performance and progressive social relations.

Conclusion

In questioning the contribution of studies of geographies of materialities and assemblage to critical urban theory, Brenner et al. (2011: 243) argue that human and non-human interactions are 'highly polysemic and promiscuous. Graffiti paint, unadorned brick, dirt in the backyard, gardens, corrugated metal ... [edit] each can be seen as an expression of precarious impoverishment of dominating, aestheticized, prosperity'. Brenner et al. go on to suggest that while meanings are transformed by context and again by the 'context of context' ... [and] the hegemony of scale ... [he challenges us to ask] how are these districts, fragments and scripts, compounds, escarpments and waterfronts must be seen and understood within the broader social spatial field of the formal city. While we broadly agree with the need to work towards such a goal, by focusing on the wearing of 'comfy clothes' in Petržalka we nonetheless follow Dovey's (2011) contention that assemblage thinking is more complex than finding root causes. We therefore support McFarlane's assertions (2011: 381–382) that 'the possibility of what *can* be understood by paying close attention to what happens at particular sites, before skipping over to pre-given analytical frames that might encapsulate something 'bigger' ... [and allowing] a focus on everyday materialities of urban sites can provide insights into the nature of poverty, inequality and urban political economies'. As such while there is more work to be done on understanding geographies of wearing of comfy clothes and urban life in specific spaces and places in cities like Shanghai, New York, Middleborough and Cardiff, we agree with McFarlane (2011: 383) that it is important 'not to lose a focus on tendencies [which] occur across multiple space-times. Here, assemblage thinking locates causality not in wider underlying contexts but *within* particular contexts [edit]... understanding multiple sites allows is to see how the ordering of urban life operates across differences and enables certain possibilities over others'.

Our findings from Petržalka are thus underpinned by the understanding that that 'context of contexts' is relational and that geographies of wearing of 'comfy' clothing in public space makes a generative contribution, for example, to 'political unity through difference' (McFarlane 2011: 385). As such, while Amin's (2008) view would suggest that the wearing of 'comfy' clothing in urban public space is 'politically modest' our research into the changing political, economic, social, cultural and spatial 'context' of the materialities and assemblages of informality and 'comfort' in Petržalka highlights the complex ways that 'practices, subjectivities and forms of identification do not *necessarily* resist ... (although they may under certain circumstances) but are attempts – sometimes unsuccessful ones – to find ways to make material life more tolerable' (Smith and Rochovska 2007: 1175). As Simone (2011b: 356–359) suggests:

> the city may be the familiar form, but it is also a ruse ... urban life is more a matter of what can be made relatable at any point in time, what can transverse established notions of the 'near and far' or 'here and there'; mobilities that leave in their wake a fabric of uneven concentrations of capacity and opportunity. In landscapes of vast inequality, of enforced conjunctions and detachments – choreographed by a variegated capitalism – life is also something rigged together from whatever is at hand.

Focusing on geographies of comfort, fashion and clothing in Petržalka, and cities and urban spaces and places elsewhere, thus offers much towards 'surfacing in plain view' changing materialities, assemblages and 'context' bound up with political, economic, social and cultural geographies of urban life. By working at the intersection of theories of fashion, materialities and assemblage we have shown how a focus on individual and collective practices thus offers fruitful insights into the ways in which people claim a material right to the city in a manner that enriches both critical urban theory and foregrounds fruitful avenues of research in studying geographies of comfort, consumption and consumer culture.

Notes

1 This chapter is a shorter version of Jayne, M. and Ferenčuhová, S. (2013) 'Comfort, identity and fashion in the post-socialist city: Materialities, assemblages and context', *Journal of Consumer Culture*, 15(3), 329–350.

2 This chapter is based on ethnographic evidence from participant observation in public/private spaces and 15 in-depth interviews with Petržalka residents, as well as five with residents from elsewhere in Bratislava. To ensure anonymity respondents have been allocated pseudonyms. Interviews were recorded and were analysed using conventional qualitative techniques.

References

Amin, A. (2008) 'Collective culture and urban public space', *City*, 12(1), 5–24.

Amin, A. (2002) 'Spatialities of globalization', *Environment and Planning A*, 34, 385–399.

BBC News (2011) *Head teacher appeal to school run 'pyjama parents'*, http:/www.bbc.co.uk/news/uk-engalnd-tees-13511668

Benjamin, W. (1982) *The arcades project.* Harvard University Press: Harvard.

Bernard, M. (1996) *Fashion as communication.* Routledge: London.

Bitusikova, A. (1998) 'Transformations of a City centre in the light of ideologies: the case of banska Bystrica, Slovakia', *International Journal of Urban and Regional Research*, 22(4), 614–622.

Borden, I. (2001) *Skateboarding, space and the city: architecture and the body.* Berg: London.

Bovone, L. (2006) 'Urban style cultures and urban cultural production in Milan: postmodern identity and the transformation of fashion', *Poetics*, 34, 370–382.

Brenner, N. (2009) 'What is critical urban theory?' *City*, 13(2–3), 198–207.

Brenner, N., Madden, D.J. and Wachsmuth, D. (2011) 'Assemblage urbanism and the challenges of critical urban theory', *City*, 15(2), 225–240.

Breward, C., Gilbert, D. and Lister, J. (2006) *Swinging sixties: fashion in London and beyond 1955–1970.* V&A Publications: London.

Bromley, N. (2004) 'Flowers in the bathtub: boundary crossings at the public-private divide', *Geoforum*, 36, 281–296.

Cannon, A. (1998) 'The cultural and historical contexts of fashion', in Brydon, A. and Niessen, S. (eds) *Consuming fashion: adorning the transnational body.* Berg: Oxford, 23–34.

Colls, R. (2004) 'Looking alright, feeling alright: emotions, sizing and geographies of women's experiences of clothing consumption', *Social and Cultural Geography*, 5, 583–596.

Czepczynski, M. (2008) *Cultural landscapes of post-socialist cities.* Ashgate: Aldershot.

De Boeck, F. (2011) 'Spectral Kinshasa: building the city through architecture of words', in Edensor, T. and Jayne, M. (eds) *Urban Theory beyond the west: a world of cities.* Routledge: London, 311–328.

Deleuze, G. and Guattari, F. (1988) *A thousand plateaus*, trans. B Massumi. London: Atlhone Press.

Dovey, K. (2011) 'Uprooting critical urbanism', *City*, 15(3–4), 347–354.

Entwistle, J. (2000) *The fashioned body: fashion, dress and modern social theory.* Polity Press: Cambridge.

Farias, I. (2011) 'The politics of urban assemblages', *City*, 15(3–4), 365–374.

Farias, I. and Bender, T. (2010) *Urban assemblages: how actor network theory has change urban studies.* Routledge: London.

Ferenčuhová, S. (2011) 'Urban theory beyond the "East/West divide"? Cities and urban research in postsocialist Europe', in Edensor, T. and Jayne, M. (eds) *Urban theory beyond the west: A world of cities.* Routledge: London, 65–74.

Ferenčuhová, S. (2011) 'Coping with the past': Urban planning in a post-socialist city', in Eckardt, F. and Elander, I. (eds) *Urban Governance in Europe.* BWV, Berliner Wissenschafts-Verlag: Berlin, 275–296.

Ferenčuhová, S. and Jayne, M. (2014) 'Zvyknúť si na Petržalku: každodenný život, bežná spotreba a vzťah k socialistickému sídlisku' (Getting used to Petrzalka: everyday life, everyday consumption, in a socialist housing estate), *Cesky Lid,* 100(3), 303–318.

Hardt, M. and Negri, A. (2009) *Commonwealth.* Harvard University Press: Cambridge, MA.

Harvey, D. (2008) 'The right to the city', *New Left Review,* 53, 23–40.

Hedidge, D. (1979) *Hiding the light: on images and things.* Routledge: London.

Holliday, R. (1999) 'The comfort of identity', *Sexualities,* 2(4), 475–491.

Horschelmann, K. and Stenning, A. (2008) 'Ethnographies of postsocialist change', *Progress in Human Geography,* 32(2), 339–361.

Iossifova, D. (2011) 'Shanghai borderlands: the rise of a new urbanity?', in Edensor, T. and Jayne, M. (eds) *Urban theory beyond the west: a world of cities.* Routledge: London, 195–208.

Irish Central (2012) *Jobseekers banned from wearing pyjamas at Dublin dole office,* www.irishcentral.com accessed 17 March 2012.

Latham, D. and McCormack, D. (2004) 'Moving cities: rethinking he materialities of urban geographies', *Progress in Human Geography,* 28(6), 701–724.

Latour, D. (2005) *Reassembling the social: an introduction to actor–network theory.* Oxford University Press: Oxford.

Lefebvre, H. (2009 [1966]) 'Theoretical problems of autogestion', in Brenner, N. and Elden, D. (eds) *State, space, world.* University of Minnesota Press: Minneapolis, 138–152.

Lefebvre, H. (1996 [1968]) 'The right to the city', in Kofman, E. and Lebas, E. (eds) *Writing on the cities.* Blackwell: Cambridge, MA, 63–184.

Maffesoli, M. (1994) *The time of tribes: the decline of individualism in mass society.* Sage: London.

Manjoo, F (2012) *The pyjama manifesto,* www.slate.com accessed 28 March 2012.

McCann, E. and Ward, K. (2011) *Mobile urbanism: city policymaking in the global age.* University of Minnesota Press: Minnesota.

McFarlane, C. (2011) '"On context": assemblage, political economy and structure', *City,* 15(3–4), 375–388.

McFarlane, C. (2009) 'Translocal assemblages: space, power and social movements', *Geoforum,* 40, 561–567.

Miller, D. (2008) *The comfort of things.* Berg: London.

Moran, L. and Skeggs, M. (2004) S*exuality and the politics of violence and safety.* Routledge: London.

Mort, F.C. (1996) *Cultures of consumption: Masculinities and social space in late-twentieth century Britain.* Routledge: London.

Rao, V., De Boeck, F. and Simone, A. (2007) 'Invisible urbanism in Africa', *Perspecta, 39,* 78–91.

Schmid, C. (2012) 'Henri Lefebvre, the right to the city, and the new metropolitan mainstream', in Brenner, N., Marcuse, P. and Mayer, M (eds) *Cities for people, not for profit: critical urban theory and the right to the city,* Routledge: London, 63–85.

Sennett, R. (2003) *Respect. The formation of character in an age of inequality.* Penguin: London.
Shove, E. (2003) *Comfort, cleanliness and convenience: The social organization of normality.* Berg: London.
Simmel, G. (1957) 'Fashion', *American Journal of Sociology*, 62(6), 541–558.
Simone, A. (2011a) 'The surfacing of urban life: a response to Colin McFarlane and Neil Brenner, David Madden and David Waschmuth', *City*, 15(3–4), 355–365.
Simone, A. (2011b) 'No longer the subaltern: refiguring cities of the global south', in Edensor, T. and Jayne, M. (eds) *Urban theory beyond the west: A world of cities.* Routledge: London, 31–46.
Slovakian National Statistics (2010) http://px-web.statistics.sk/PXWebSlovak/index.htm accessed 12 December.
Smith, A. (2002) 'Culture/economy and spaces of economic practice: positioning households in post-communism', *Transactions of the Institute of British Geographers*, 27, 232–250.
Smith, A. and Rochovska, A. (2007) 'Domesticating neo-liberalism: everyday lives and the geographies of post-socialist transformation', *Geoforum*, 38, 1163–1178.
Stenning, A. (2000) 'Post-socialism and the changing geographies of the everyday in Poland', *Transactions of the Institute of British Geographers*, 30, 113–127.Stenning, A., Smith, A., Rochovská, A. and Świątek, D. (2010) 'Credit, debt, and everyday financial practices: Low-income households in two postsocialist cities', *Economic Geography*, 86(2), 119–145.
Swain, J. (2002) 'The right stuff: fashioning an identity through clothing in a junior school', *Gender and Education*, 14(1), 53–69.
Veblen, T. (1898) *The theory of the leisure classes.* Penguin: London.
Wilson, E. (1985) *Adorned in dreams: fashion and modernity.* London Virago.
Zamperini, P. (2003) 'On their dress they wore a body: fashion and identity in Late Qing Shanghai Positions', *Positions: East Asia Cultures Critique*, 11(2), 301–303.

11 Cosy, comforting, disruptive?

Knitting and knitters in/out of place

Laura Price

Comfy clothing and familiar fabrics

Textile historian Beverly Gordon (2011: 146) suggests that fundamentally "everyone, in every culture, uses and understands cloth; everyone has kinetic experience with fabric and its comforting properties". However, this notion of comfort, of cloth being comforting, is much more than just the *feel* of fabric – that is to say, the embodied experience of wearing certain clothing. Certain fabrics and therefore certain clothing: jeans, woollen jumpers, knitted cardigans, for example, are, in the Global North at least, considered more everyday, ordinary and vernacular than others. As Jo Turney (2014: 22) notes, knitting is "familiar and familial, it is the stuff of everyday life". But, indeed, how do certain fabrics and materials become associated with comfort, and the doing of comforting?

On comfort, Daniel Miller and Sophie Woodward (2012: 65) suggest it is a "physical experience that encodes a sense of what seems suitable or appropriate for a particular person". Through the example of blue jeans (part of their broader work into global denim and clothing systems) they suggest "jeans have the capacity to make people feel more comfortable than other garments because they can elide the physical idea of comfort with the social concern of how the wearer looks in public" (Miller and Woodward 2012: 77). With that in mind, are there certain fabrics and clothes that can be *too* comfy, too comfortable to wear publicly? As this volume explores, comfort oscillates between comfort and discomfort – it is a liminal space. For example, pajamas or sportswear are often thought as only being acceptable to wear "at home" despite it being an important feature of public life around the world (Jayne and Ferenčuhová 2015: 335). Comfort might be read as embodying *resistance* to the hegemonic discourses of "proper" feminine behaviour and attire (Holliday 1999: 479).

Moreover, such historically "feminine" materials that have "strong tactile associations, such as wool, hair, silk, feathers, shells and beads" (Classen 2012: 133) have often been neglected by contemporary social science – such *wooliness* is not taken seriously in every material, social and political sense (Price 2015). In this chapter I locate my research within rich research into women's work, textile-making and the history of cloth – to explore knitting as a vernacular creative practice and familiar material process. In doing so, I discuss the spatial politics of comfort that have historically associated this craft with comfort, coziness and therapeutic space – that through spatial politics and interventions in public/private boundaries becomes disruptive and political. I illustrate this through ethnographic research and interviews conducted with amateur knitters and knitting groups in London and North West England during 2012–2015.

Cultural geographers Chantel Carr and Chris Gibson (2016: 309) argue that "making is central to who we are as individuals – what we make as part of everyday practice forms our identities and place in the world". As such, geographical debate and engagement with materials, making, creativity is burgeoning, paying attention to not only the embodied experience of craft and crafting, but the social and material relations of labour, technique and skill (Price and Hawkins 2018). The chapter is located within these engagements, but more specifically *feminist* geographies of making, materials and textiles (Crewe 2011; Stanes and Gibson 2017). Feminist geographers have brought to attention creative practices that are neglected because of their gendered connotations and the "softer, sensory, emotional, and tactile characteristics more typically associated with dress" (Crewe 2011: 2094). Throughout this chapter I explore the implications and possibilities of engaging with the texture and vibrancy of yarn, wools and fibre; and explore their role in generating and disrupting comfort. In doing so, I engage with Sara Ahmed's work on comfort, the spatial politics of comfort and the notion of "public comfort" (Ahmed 2008, 2010).

> I didn't realise it's so relaxing, just the wool on the needles.

Careful knitters is a weekly knitting group based in my "hometown" of Warrington, in the North West of England. The knitters meet weekly in a local community centre and for four hours during the daytime, midweek. All members are women and all managed, in various capacities, caring responsibilities supporting family members with their physical and mental health. The group is designed to create a space for its members to have time to relax, through shared activities and a social network that offers some respite from the unpaid caring work they undertake each week.

The group often worked individually toward collective goals, such as yarnbombs (a form of knitted graffiti, for example a wrapping around a tree) and the production of knitted items that provide assistance and comfort to hospital patients. For example, twiddlemuffs – a double thickness hand muff with embellishments attached inside and out. It is designed to provide a stimulation activity for restless hands for patients suffering from dementia. It was the group leader, Sarah, that introduced knitting as an activity "to help bring everyone together, but also to help everyone relax". These sentiments echo research that illustrates the generative effects of knitting on well-being and mental health (Riley 2013; Corkhill *et al.* 2014) – through the rhythm of making, and feeling of flow.

Geographers have begun to take seriously the potential of vernacular creativity for well-being in communal, individual, and embodied sense. (Pitt 2015; Price and Hawkins 2018). Knitting in a group is reported to improve social confidence and *feelings of belonging* (Corkhill *et al.* 2014). One of the respondents of the survey into the benefits of knitting to mental health reported "When I knit during situations where I would normally feel incredibly uncomfortable and anxious, I feel less so. I'm able to enjoy the fact that I am around people and it is always a nice icebreaker" (Corkhill *et al.* 2014: 38). Indeed, the process of knitting was described by respondents as "soothing", "restful", or spiritual" with the "rhythm of the repetitive motion" being "hypnotic" and "calming" (Corkhill *et al.* 2014: 37). This was echoed by one member from Careful Knitters, who reinforced such notions:

> It gets to that stage, where I'm knitting before my husband's got in, and I think I've got to get his tea on, but I don't want to put my knitting down. (…) I didn't realise it was so comforting really, just the wool on the needles and how it's growing like. I must admit I get carried away; the time just goes so quick, you're not thinking of anything else, you're just relaxing.
>
> (Barbara, Interview, November, 2013)

For Sara Ahmed (2010: 156),

> comfort is about an encounter between more than one body, which is the promise of a "sinking" feeling. To be comfortable is to be so at ease with one's environment that it is hard to distinguish where one's body ends and the world begins.

Barbara gets "carried away" by her knitting: "carried away" to a different sense of time and space, where tea can wait and time flies by; "carried away" into an interior state of relaxed, comforting focus; "carried away" with her own pleasure. The domestic space of social relations and obligations and work becomes a space-time of self-preoccupation and nurture. Ciara echoes this sentiment in talking about the good that knitting does her mother. Ciara is a member of Careful Knitters – she attends each week with her mother, Sarah:

> Well yeah, I think it's for my Mum really. It's time for herself. She does all the dealing with my Dad, all his medication and everything. She gets stressed out. But something like this knitting group, because it's like ten 'til two [on a week day] it's probably like "chill time" here. It's just to get away really isn't it? – then you go back home and you do what you're doing and you're ready to get back to it.
>
> (Ciara, Interview, November, 2013)

This is not to suggest that Sarah is *un*comfortable at home. However, home is a place of significant emotional and physical labour. Through multiple caring responsibilities Sarah may not frequently be able to experience corporeal feeling of relaxation and comfort, as described by David Bissell (2008: 1697) as affectual circulations that flow through the sedentary body induced in relation to proximate objects and things. That said, feminist geographers, particularly, have sought to counter the home as haven thesis, as a space of rest and safety highlighting, rather home has historically been a space of emotional and physical labour (paid and unpaid) and inequalities perpetuate for whom home is a place of comfort (See Brickell 2013).

Yet the notion that "home" should be a space of contentment, or indeed "the comfort zone", is pervasive in public spatial imaginary. Narratives of home are closely intertwined with spaces of comfort, of ease, well-being and satisfaction (Holliday 1999; Moran and Skeggs 2004; Ahmed 2006). Ciara highlights how the group is generative of comforting sensations and well-being for her mother. Not only through the embodied act of knitting which is physically, viscerally, affectively relaxing, but rather the space of the knitting group itself *produces* comfort – generated by the material and social relations that knitting affords. It is carved out as "chill time". As Boulila (2015: 135) writes, practices that create comfort can therefore be conceptualised as spatial technologies that enable subjects to extend into spaces. In the case of this weekly knit group it was achieved in part through its familiarity in terms of temporality (10 am till 2 pm on the same weekday), but more importantly through the particular sociability afforded through making, doing and crafting (Price and Hawkins 2018).

Knitting in/out of place: public comfort

In this part of the chapter I now want to turn to explore how knitting groups can *disrupt* the spatial "difficult and slippery politics of comfort" in public and private spaces (Moran and Skeggs 2004: 103). Bodies that knit materially occupy space in multiple ways, by being active, or becoming, in situ. Not only this, but projects and matter are becoming too in the making process (Ravetz *et al.* 2013). Knitting groups take up a sizeable amount of room – arms knitting with two needles, people spinning, yarns and needles being passed to and fro. Feminist geographers have long acknowledged the ways that women's bodies and materials become "out of place": "the idea that certain bodies 'get in the way' (either materially, symbolically or both) disrupting the comfort of others" (Boyer 2011: 552). Building upon the work of Sara Ahmed, Boyer (2011: 556) notes "the gendered work of maintaining public comfort"; and, through the example of breastfeeding, highlights how mothers disrupt the comfort of others by performing a domestic act that many deem should take place indoors, at home, out of public view.

The history of knitting and politics of mothering are deeply entwined (indeed, more metaphorically Jo Turney (2012: 30) suggests "knitting is like an umbilical cord, a continuous thread linking mother to child") and often placed in the domestic and private sphere. The gendering of knitting as a craft is intimately linked to its spatial positioning as a traditionally home-based craft: "as a branch of women's work it was considered essentially domestic, undertaken within the home for the home and of value nowhere else" (Classen 2012: 133) Indeed, textile craft historically "was associated with the tactile character of other domestic labours: sewing, cleaning and caring for the family" (Classen 2012: 133). Contemporarily knitting is often practised publicly via knitting groups, craft events and festivals, and often on public transport to pass time. Indeed, every year June 18th is *World Wide Knit in Public Day*. Bratich and Brush (2011: 236) argue that "knitting in public turns the interiority of the domestic outward, exposing that which exists within enclosures, through invisibility and through unpaid labour: the production of home life". So then, knitters, with their yarns, needles and making visible the "labours of making" draw attention to a particular set of cultural practices that have historically been productive of, and associated with, home. Therefore, it can be deemed to be *out of place* in the public arena. As Myzelev (2009: 161) notes:

> similar to other activities such as embroidering, crocheting, and breastfeeding, knitting allows women and in some instances men to bring their private hobbies to public spaces and thus reformulate even if temporarily the function of public areas such as cafes, buses, and libraries.

For example, as part of my research, I visited a weekly knitting group in West London, the group met in a local café on a busy high-street and dissimilar to the North West knitting group, met primarily because each member was an amateur knitter who wanted to craft socially. Often, the knitting group would be stopped by a passer-by taking interest as I illustrate through fieldnotes:

> There's quite a big group knitting tonight, and we're taking up a sizeable amount of space in the café. Because of this, there's been a lot of encounters and interactions with other patrons as they've negotiated around our table, squeezing themselves in to fit past our group and often making comments on our knitting, sharing quick stories, or just smiling/sharing humorous glances at the situation. About half way through the group, a man squeezes past the table and says, "When do you think my pullover will be ready girls? I'm expecting it soon". His female companion tuts and nudges him as if to say "shut up". Members of the knitting group roll their eyes, and Harriet leans in and whispers, "I loathe men like that".
>
> (Hammerknit Fieldnotes, January 2013)

Alongside the micro-geographies of renegotiating space when knitting publicly; the anecdote reveals the way that knitting outside of the home disrupts the everyday by producing encounters that surprise, intrigue, excite and make laugh those who are view knitting *as out of place*. Geographers have drawn attention to the "emotional, embodied and affective practices that counter boundaries between public/private space, individual/collective experience and comfort/discomfort with reference to distance/proximity, pleasure/pain and mobilities/stillness" (Jayne and Leung 2014: 265). Ahmed (2010: 548) suggests that "maintaining public comfort requires certain bodies to 'go along with it'. To refuse the place in which you are placed is to be seen as causing trouble".

Throughout my research, I found that knitting has been harnessed as both implicitly and explicitly activist. For example, producing knitted items specifically to support activism and activist aims through craft and explicitly as part of feminist politics (see Greer 2014). However, through it was the quiet politics of implicit activisms through knitting that I found especially generative, disruptive and nuanced (Horton and Kraftl 2009; Askins 2015). Throughout my research rarely did knitters articulate *explicitly* feminist motivations for knitting publicly. However, our conversations were more often than not shaped by, and framed in, understandings of women's experience in space and place and a commitment to ways it might be otherwise.

For example:

> Anna again comments how she likes knitting because she can knit on the tube and not take up too much room. Ella says the same – she's started knitting with circular needles even for flat knitting because it takes up less room. She says with big knitting needles she worries she'll be like "one of the men who feel the need to straddle, and take up excessive room". Carrie says how important it is that we stand our ground and make men like that "know it's not okay to take up the space, but take up the space with our knitting."
>
> (Hammerknit Fieldnotes, March, 2013)

Knitting the comfort zone?

For some women, then, knitting in public is a way to claim space, to counter and subversively mimic the occupations of public space made by men, in doing so, creatively intervening into the patriarchal social and material systems of urban space (Mott and Roberts 2014). The collective voicing of affirmation recorded in the fieldnotes excerpt above is suggestive of a collectively felt sense of empowerment through the knitting group's public presence. However, not all knitters felt this way and whilst I am keen to avoid generalisations over experience of space, place and geographies it was notable that urban-dwelling knitters felt largely more comfortable to be present in public space with their knitting. However, for some, this experience may have served for them to feel *more* out of place, or "affect aliens" to use Sara Ahmed's (2010) term, that is "to experience alien affects – to be out of line with the public mood."

Not all knitters feel comfortable knitting "in public", and often this discomfort is felt through different, diverse identities. Taylor (2011: 15) suggests

> individuals do not arrive, completely dis-located from space, but rather inhabit and rework space on the basis of past dispositions, knowingness, confidences and capitals. Classed (dis)comforts and resources are significant to the uptake of, inclusion into and exclusion from space, whether that be commercialised leisure space or everyday home space.

Knitting publicly calls into question women's subjective "sense of place" and for some members of *Careful Knitters*, it would be a source of discomfort.

> Well, no I don't [knit publicly] and I don't know whether I would feel uncomfortable. It's doing something else doing that. Knitting is a pleasure, it's not an obsession, and so I don't want to go overboard with it. I enjoy doing it at home in my own space.
>
> (Shelia, Interview, December, 2013)

> I love it, but I only knit at home and here [with *Careful Knitters*]. Nowhere else. I just enjoy relaxing and doing the knit. I feel comfortable here.
>
> (Barbara, Interview, November 2013)

As one of the interviewee's noted "people don't want to step out of their comfort zone. I mean it's not only knitting, and doing something a bit different, but it's also with a group of people you don't know – so it's a double thing".

LAURA: What's it like then? Being part of the group?

CIARA: Well I don't know. Because sometimes it's like, you associate knitting with older people. And I'm not sure I'm a knitter. But I feel proud of something I've finished and I want to go on Facebook and share it but it's like "Oh, shall I do this, or not?"

LAURA: Would you not post it to Facebook then, you know, go public with your knitting?

CIARA: Well, yeah, I have done the odd time and people have been dead nice, like "wow have you made that?' But it's like, I'm only twenty, so it's a bit embarrassing for me really, my boyfriend laughs at me. I went to his house the other day and I was like 'I'm just gonna do a bit of this Twiddlemuff', and he said to me 'What next, are you gonna have a broken hip or something?' It is a good thing that I can take my knitting out and about. But I couldn't knit it in public, no. I dunno, it's just one of those things, I'm young and I can't be seen knitting – obviously I take it to my boyfriend's. But no, I couldn't. It's not cool enough. If anyone saw me, I'd be like that [covers face] – no I couldn't do that."

(Ciara, Interview, November, 2013)

In both the interview and informal conversations with Ciara she expressed how important being part of the knitting group was to her. It enabled her to feel productive, useful and needed – particularly as she searched for full-time employment. For Ciara, the knitting group was a safe, empowering space; compared to the potential vulnerability she occasionally felt knitting outside of the group. However, Ciara would not like to be seen knitting "in public".

This sentiment was shared by other knitters, like Shelia, who knit only at home and in the company of her family or friends for fear of her enthusiasm being defined as "obsessive". As Geoghegan (2013: 45) notes, "whilst enthusiasm as an emotional affiliation gives rise to senses of self and feelings of belonging and attachment, it also has the potential to disrupt, challenge and alter social relations in space". Ahmed (2010: 175) suggests "to become uncomfortable, means becoming not comfortable" she adds, "in other words, the intensification of affect is what is noticeable: certain affects can hover in the background as your affective situation, your 'around' or surround, which comes to your attention through the accumulation of intensity" (Ahmed 2010: 175). I would argue that Ciara frames the notion of knitting publicly, not as necessarily *uncomfortable*, but rather it would be lacking familiarity and security that knitting within the group provides. It would be indicative of a level of sociability and knowledgeability of knitting that to her non-knitting peers, she fears would be a show of "too much" enthusiasm (Geoghegan 2013). The therapeutic experience of knitting could potentially be opened up to public judgements of Ciara's crafting identity and that spaces feel insecure, and unsafe. So then, whilst for some knitters the disruption of comfort empowers and performs a feminist re-coding of public space through the potential disruption of public comfort (Ahmed 2010); for others, the empowerment of knitting comes from creating a safe space outside of everyday public gazes. The comforting affects are felt as therapeutic by their temporality, time and space outside of ordinary, everyday life.

Members of *Hammerknit* were not only confident and empowered by knitting publicly, but it could be argued that their knitting had more explicitly political geographies – based around challenging perceptions of craft, women who knit, and expectations of where knitting should take place. As mentioned above in my *Hammerknit* Fieldnotes (March 2013), it is important for some knitters to visibly "take up space" with their knitting. This is a reminder that the felt bodily comfort and performance of knitting, the affects and emotions it circulates, are not devoid of the powerful, complex politics of the comfort of self, identity and place. Rather, it is precisely in the spaces that these registers of social, public, private and bodily comfort collide and oscillate, that feminist creative geographers should be attuned to (Holliday 1999; Moran and Skeggs 2004).

The sense of empowerment or disempowerment that knitters I researched felt through knitting publicly is, I argue, inextricably bound up with broader discourses of identities and life beyond the knitting group. This is not to belie the powerful world-making potential of knitting and knitting groups. Such spaces support and generate the feeling of flow, and comforting sensations produced through making, doing, and social interaction under the guise of knitting with knitters. However, on occasion, the affective energies of the knitting group are compromised and re-navigated. Such affectual route changes, I argue, occur when broader social identities and personal geographies come back to the fore and are negotiated as such.

For some knitters, knitting stays at home, or in quasi-public-private spaces such as the community centre that hosts their weekly knitting group. In doing so, such interactions complicated and interrogated historical narratives and assumptions around home as a place of comfort, or rather that home is where comfort takes place (Moran and Skeggs 2004). For some knitters, knitting in public is a form of both explicit and implicit craftivism aligned with broader feminist concerns of bodies, space and the city (Horton and Kraftl 2009; Askins 2015). Such knitters felt empowered by making their creative practice public and disrupting expectations of home, place and comfort (Ahmed 2010). For others it is more emotionally, physically and mentally rewarding if their knitting stays at home or in specific times in quasi-public-private spaces such as community centres. Such narratives both align with and invite nuance into contemporary engagements with craft as a tool of social connection (Price and Hawkins 2018; Guantlett 2011).

In many ways it is an unsurprising argument to suggest that some knitters disrupting comfort and exploring discomfort is a tool of empowerment and substance. Not least because many have argued that societal change, personal growth and discovery occurs through discomfort, rather than comfort – such change is not supposed to *feel* cosy as it protests and intervenes with entrenched narratives and systems that seek to uphold public comfort (Ahmed 2010). Whilst for some the act of knitting publicly is an empowering pastime, and one that can be used to challenge normative assumptions about public comfort and private/public boundaries, for others such potentially political geographies are not a comfortable experience, as shown through Ciara's reflection: "I couldn't knit it in public, no. I dunno, it's just one of those things, I'm young and I can't be seen knitting".

It is too simplistic to see this as a matter of personal confidence; rather, this reflects how the identifications and affectual forces of being a knitter, and therapeutic material relations therein, intersect with other aspects of identity, and thus how they are embodied and felt. The spatial relationships and ideas of mobility/immobility worked through knitting groups via notions of comfort/discomfort are complicated, complex and worthy of future research. Not least, as geographical engagement with textiles, materials, craft and domestic skill is undervalued (Stanes and Gibson 2017). Public knitting is not only gendered in my research, but it is shaped by age and class too. Not everyone who knits moves (or wants to move) their knitting into public spaces; there are complex spaces of domesticity and notions of femininity to which people feel subscribed (Appleford 2016).

Conclusion

This chapter has explored the geographies, registers and spaces of comfort that knitting affords by investigating the affects that are produced and circulated by the embodied doing of knitting and being part of knitting groups. Firstly, by beginning with the space of the body, I suggest that knitting has therapeutic potential and bodily experiences. Knitting is bounded up with notions of domesticity, home and comfort – though feminist geographers seek to trouble such easy associations and narratives. Through my engagement with knitters for whom home does not always present a place of rest, I illustrated the productive power of knitting and knitting groups as a space of comfort, safety and relaxation outside of the home.

Secondly, I caution to recommend that such a sense of well-being and relaxation can be found by all in the knitting group. Rather, this depends on a sense of comfort in space which moves beyond the affectual, bodily sensation of relaxing but rather attends to the wider comfort of subjectivities and identities – in this case, especially around age and classed identities. Thirdly, I introduce the complex ways that historically "homely" narratives around traditionally women's creative practices, such as knitting, can intervene and challenge normative senses of public comfort (Ahmed 2010). This is not a linear or dichotomous process, but rather the complex politics of comfort/discomfort are harnessed, or avoided, by knitters with diverse subjectivities and identities.

To conclude, paying attention to the geographies of comfort in spaces of craft and making allows for nuanced engagement of the spatial politics of bodies, practices and spaces therein. Future work on textile materials and making might explore the way that public displays of amateur creative practices successfully intervene, subvert and re-imagine space differently. Or rather, reinforce boundaries, places and identity. This chapter modestly contributes in quiet and disruptive ways to geographical engagements with the historically gendered craft of knitting and the complex ways in which this cosy and comforting practice disrupts and interrogates personal, private and public worlds.

Bibliography

Ahmed, S. (2006) *Queer Phenomenology*. Durham: Duke University Press.
Ahmed, S. (2008) "Sociable happiness." *Emotion, Space and Society*, 1(1), 10–13.
Ahmed, S. (2010) *The Promise of Happiness*. Durham: Duke University Press.

Appleford, K. (2016) "Being seen in your pyjamas: The relationship between fashion, class, gender and space." *Gender, Place & Culture*, 23(2), 162–180.

Askins, K. (2015) "Being together: Everyday geographies and the quiet politics of belonging." *ACME: An International Journal for Critical Geographies*, 14(2), 470–478.

Bissell, D. (2008) "Comfortable bodies: Sedentary affects." *Environment and Planning A: Economy and Space*, 40(7), 1697–1712.

Boulila, S. C. (2015) "What makes lesbian salsa space comfortable? Reconceptualising safety and homophobia." In K. Browne and E. Ferreira (eds) *Lesbian Geographies: Gender, Place and Power.* London: Routledge, pp. 149–168.

Boyer, K. (2011) 'The way to break the taboo is to do the taboo thing' breastfeeding in public and citizen-activism in the UK. *Health & Place*, 17(2), 430–437.

Bratich, Z. J., & Brush, M. H. (2011) "Fabricating activism: Craft-work, popular culture, gender." *Utopian Studies*, 22, 234–260.

Brickell, K. (2013) "Towards geographies of speech: Proverbial utterances of home in contemporary Vietnam." *Transactions of the Institute of British Geographers*, 38(2), 207–220.

Carr, C., & Gibson, C. (2016) "Geographies of making: Rethinking materials and skills for volatile futures." *Progress in Human Geography*, 40(3), 297–315.

Classen, C. (2012) *The Deepest Sense: A Cultural History of Touch.* Champaign: University of Illinois Press

Corkhill, B., Hemmings, J., Maddock, A., & Riley, J. (2014) "Knitting and well-being." *Textile*, 12(1), 34–57.

Crewe, L. (2011) "Life itemised: Lists, loss, unexpected significance, and the enduring geographies of discard." *Environment and Planning D*, 29(1), 27–46.

Geoghegan, H. (2013) "Emotional geographies of enthusiasm: Belonging to the telecommunications heritage group." *Area*, 45(1), 40–46.

Gordon, B. (2011) *Textiles: The Whole Story.* London: Thames and Hudson.

Greer, B. (2014) *Craftivism: The Art of Craft and Activism.* Vancouver: Arsenal Pulp Press.

Guantlett, D. (2011) *Making Is Connecting: The Social Meaning of Creativity, from DIY and Knitting to YouTube and Web 2.0.* Cambridge: Polity Press.

Holliday, R. (1999) "The comfort of identity." *Sexualities*, 2(4), 475–491.

Horton, J., & Kraftl, P. (2009) "What (else) matters? Policy contexts, emotional geographies." *Environment and Planning A: Economy and Space*, 41(12), 2984–3002.

Jayne, M., & Ferenčuhová, S. (2015) "Comfort, identity and fashion in the post-socialist city: Materialities, assemblages and context." *Journal of Consumer Culture*, 15(3), 329–350.

Jayne, M., & Leung, H. H. (2014) "Embodying Chinese urbanism." *Area*, 46, 256–267.

Miller, D., & Woodward, S. (2012) *Blue Jeans: The Art of the Ordinary.* Berkley: University of California Press.

Moran, L. J., & Skeggs, B. (2004) *Sexuality and the Politics of Violence and Safety.* London: Routledge.

Mott, C., & Roberts, S. M. (2014) "Not everyone has (the) balls: Urban exploration and the persistence of masculinist geography." *Antipode*, 46, 229–245.

Myzelev, A. (2009) "Whip your hobby into shape: Knitting, feminism and construction of gender." *Textile*, 7(2), 148–163.

Pitt, H. (2015) "On showing and being shown plants – a guide to methods for more-than-human geography." *Area*, 47(1), 48–55.
Price, L. (2015) "Knitting and the city." *Geography Compass*, 9(2), 81–95.
Price, L., & Hawkins, H. (eds) (2018) *Geographies of Making.* London: Routledge.
Ravetz, A., Kettle, A., & Felcey, H. (2013) *Collaboration through Craft.* London: Bloomsbury.
Riley, J. (2013) "The benefits of knitting for personal and social well-being in adulthood: Findings from an international survey." *The British Journal of Occupational Therapy*, 76(2), 50–57.
Stanes, E., & Gibson, C. (2017) "Materials that linger: An embodied geography of polyester clothes." *Geoforum*, 85, 27–36.
Taylor, Y. (2011) "Sexualities and class." *Sexualities*, 14(1), 3–11.
Turney, J. (2012) "Making love with needles: Knitted objects as signs of love?" *Textile*, 10(3), 302–311.
Turney, J. (2014) "A sweater to die for: Fair Isle and fair play in the killing." *Textile*, 12(1), 18–33.

12 A correspondence with water

On the (dis)comforts of the swimming pool

Miranda Ward

i

Wales, May 2014. Here I am, missing the pool, the act of doing my daily lengths, even though I've been looking forward to this break for months. I brought my suit and my cap and my goggles with me, as I always do, but we're in a cottage in rainy Powys and the likelihood of a swim is slim: it's more of a glass-of-wine-by-the-fire kind of holiday. Still, the kit takes up almost no space in a rucksack heavy with waterproof jackets and boots and books, and it's an old habit, inherited from my dad, carried over from California: *always pack your swimsuit.*

I've also packed Olivia Laing's *To the River*, in which Laing follows the River Ouse in Sussex from its source to where it flows into the English Channel. It's not a book about swimming, but it is very waterlogged. Laing is preoccupied throughout with Virginia Woolf, who ended her life in the water, and Laing herself often bathes in the river on her journey, floating, wallowing: "I trod water in the wide pool by the rushes, leaning my head back till the silt soaked into my scalp," she writes at one point, having found a suitable spot and "plopped in" (Laing 2011: 112).

As I read I start to wonder what it is exactly that constitutes a swim. What is the act itself, the verb? By definition it connotes motion, action: *"to move through water…"* – and so it has always seemed to me that submersion is not necessarily the same thing as swimming. There's obviously shared territory here, however, and wild-swimming types, I notice, often seem to blur the boundaries completely, so that a quick dip, a float, in any conceivable body of water, is as much a form of swimming as some rigid, lengthy session in a pool.

I can see why, on one hand: fundamentally, "plopping in", letting the silty river soak into her scalp, puts Laing in contact with water, just as my up-and-down-the-lanes routine puts me in contact with water; in this sense it's as simple as that, it's about touch, immersion, communication or communion with the water itself. On the other hand, it feels like a very different endeavor – almost as if they're two distinctive species of activity, one centred on immersion in not only a body of water but that body's wider environment (trees, wind, sun, a grassy bank), the other, the pool swim, more self-oriented, centred on immersion in the swimmer's own body (arms, legs, lungs).

I know the pool is not to everyone's taste. Some people, many people, prefer the waves and the weeds. "For some outdoor swimmers," writes Ronan Foley in a study of therapeutic blue space in Ireland,

> swimming in a pool was preferable to not swimming at all, though others identified such spaces as uninteresting. In part this was to do with the artificial and non-natural nature of such spaces, characterized by chlorine and an oppressive atmosphere. For others it was precisely because of the "straight-line" predictability of the encounter and the more open possibilities of the sea.
>
> (Foley 2015: 223)

And yet the banality of the pool seems to force some reckoning with the activity of swimming itself, since the surroundings are purpose-built. I've found lately that I derive an inferior sense of satisfaction from wild submersion, as if it's holier on the surface, but also wholly on the surface. When I'm weary, when I'm stuck, when I'm hungover and fuzzy-headed, a plunge in the sea, the sting of cold water on skin, will do wonders – but I always feel a little like I've cheated, like I haven't got a real swim in, like I don't quite deserve the reward. Perhaps because it feels less like work. Perhaps because in the relative safety and predictability, in the almost religious routine – the daily pilgrimage to the pool, the devout repetitions of lengths – is a depth of meaning all its own that gets lost in the silt or salt of a wilder body of water.

ii

In thinking this through I'm reminded of Roger Deakin, who in the abstract captures so well what I find compelling about swimming as a subject of study, and yet whose recorded swims are often a world away from my own familiar efforts at the pool, as when, in the River Test, he happily glides, "downstream, brushed by fronds of water crowfoot that gave cover to trout," immersed in the landscape he observes and in this case actually propelled by the water itself (Deakin 2000: 19).

Instinctively I want to say something about this which sets it in conflict with the act of swimming laps in a pool, something about the carefree almost-laziness of it, the audacity of it, perhaps – but when it comes down to it I find only this: that there is an historically implicit masculinity in this particular genre of swim. In Charles Sprawson's *Haunts of the Black Masseur,* I remember reading of the poet Rupert Brooke that, "When working at the Bodleian he would get up in his country cottage long before dawn and bathe as he walked down to Oxford in the streams among the Cumner hills" (Sprawson 2013: 187). I was working at the Bodleian myself when I first came across this passage and it brought home to me the fact that some of my love for the pool is actually more about fear than any positive force. At the pool I feel secure enough to work, to exercise my body; at the edge of rivers I hesitate, ask, "Is it allowed? Is it safe?" I worry over questions of permission (Deakin, by contrast, resolves when confronted with the accusation of trespassing "to stand up for my rights as a free swimmer" (Deakin 2000: 31), launching himself merrily into exactly the kind of heated conversation I would run from). I worry over reports read and only half-recalled about poor water quality. I worry over what will happen to my possessions if I leave them unattended on a bank or a shore. Whereas I notice that the men – and they are mainly men – in Sprawson's (2013) romantic account of the history of swimming are always striding through landscapes, and, when the urge takes them, confidently stripping down, getting wet, communing with nature: "bathing," which sounds so innocuous, so domestic, but which seems to take on a spiritual, even divine, importance. To *bathe* in this context is not a trite act, not a sedentary, solitary soak in bubbles on a cold winter day with the radio in the background and a cup of tea balanced on the edge of the tub; it's a forceful, defiant, physical act – wild, transcendent, a way of staking a claim on the landscape, inserting oneself into it.

I have often, and perhaps ironically, thought of swimming laps as the "purest" form of swimming – just the body and the water, intimate in isolation – but it occurs to me now that it may not be so simple. There is no one true way of being in the water, no hierarchy of settings for a swim, and perhaps any distinction is not, in itself, very useful, since the substance central to both activities – water – is so slippery, so fluid. When you strip away surroundings – the trees, the tiles, the grassy bank, the leisure centre walls – what you're left with is an encounter between a body and a body of water; a communication between the two, a poetic or rhythmic connection.

iii

At one point, idly pursuing a train of thought down an Internet rabbit-hole, I encounter an entry on Wikipedia that claims that the word "to swim" comes from, amongst other sources, the Proto-Germanic *swimmaną*: "to swoon, lose consciousness, swim". I can find no other evidence that this is true, but it continues to stick with me, long after I've attempted to abandon it as a serious idea that I can incorporate into my research. It seems to have emotional if not historical veracity. To swim, to swoon, to lose consciousness: not necessarily in the sense of blacking out, but in some other, more fluid way. To lose consciousness of one way of being, of time and the body as felt on land, but then to (re)gain a new, or different, consciousness: of the body as felt in water; the body's history in water; the expansion and slippage of time. The unknown known, just at the outside edge of perception or understanding. An almost-memory, perhaps: "To be embraced and sustained by the light green water was less a pleasure, it seemed, than the resumption of a natural condition," writes John Cheever in his short story "The Swimmer", as his hero briefly considers embarking on a swim sans trunks (2009: 727).

In her book, Laing muses about the authors Iris Murdoch and John Bayley, married for over 40 years, who maintained a dedication to swimming even as Murdoch succumbed to Alzheimer's in the late 1990s. "I am not in the athletic sense a keen swimmer, but I am a devoted one," Murdoch opened her review of Sprawon's *Haunts of the Black Masseur* (Murdoch 1993). Her fiction, meanwhile, was also suffused with swimming references. "I suppose you could see the love of swimming as part of this desire to retreat or be immersed, to enter that pre-literate continuous world," writes Laing of Murdoch and Bayley; Martin Amis, she remembers, "once wrote that the couple suffered collectively from *nostalgie de la boue,* literally the desire to return to the sticky mud of one's origins, the ooze and squalor of infancy" (Laing 2011: 107).

Charles Sprawson makes the connection between swimming and "the ooze and squalor of infancy" even more confidently: "It is generally accepted," he writes, "that in the area of the unconscious, water in any form and immersion in it suggests a hidden desire for a return to the security and irresponsibility of the womb and its amniotic waters" (2013: 143). Sprawson is referring at least in part, of course, to Freud, for whom water held symbolic psychological meaning; in *The Interpretation of Dreams*, for instance, he suggests that dreams of water may be "based upon fantasies concerning the intra-uterine life, the sojourn in the mother's womb, and the act of birth" (Freud 1997: 262). The Hungarian psychoanalyst Sándor Ferenczi, a colleague of Freud's, took this symbolism further, into evolutionary, and not just gestational, depths: "What," Ferenczi posited, "if the entire intrauterine existence of the higher mammals were only a replica of the type of existence [of] that aboriginal piscine period?" (Ferenczi 1989: 45).

If this is too tenuous or fraught a claim to resolve wholly one way or another, there are nevertheless swimmers in this world for whom it is not inconceivable that the act of swimming is pleasurable – comforting – precisely because it recalls some earlier existence. Stephanie, for instance, from France, retired, in her 60s, has a theory about her devotion to the pool. "I love the feeling of the water," she tells me.

> I find it extremely relaxing. And when I come out, I feel younger. I feel it gives me energy. And – I wonder: I mean it is only my feeling, but I was born with a caesarian section, and my daughter was as well – I just wonder whether you get out of the fluid, and you have this – it gives you, it makes you really peaceful to go back into fluid. Maybe that's why I like swimming.
>
> (Stephanie, interview, August 2014)

iv

Certainly water, as both idea and element, is powerfully and variously evocative: of the womb, maybe, but also of other equally intangible and haunting things. As Ivan Illich puts it, "water has a nearly unlimited ability to carry metaphors" (1986: 24–25). It may be variously perceived as holy, as healing, or as powerful in some other not quite explicable way, and it is often linked to the very essence of life itself. The anthropologist Veronica Strang points out that hydrolatry – water worship – "occurs in every cultural and temporal context, and [...] even in the most secular cosmologies, water is presented as the fundamental source of life" (Strang 2005: 105). In a Christian context water is, of course, a crucial ingredient for baptism, symbolic of renewal. As the historian and philosopher Mircea Eliade writes, "Water purifies and regenerates because it nullifies the past, and restores – even if only for a moment – the integrity of the dawn of things" (1996: 194–195). But a tension exists between this interpretation and what Thomas Farber calls "[m]an's ever-increasing domination of water" (via, for example, "channels, drains, canals, sewers, dams, pipes, taps, hydrants, water towers, swimming pools, showers baths, sinks" (Farber 1994: 22). Such domination can be read as having stifled the spiritual or symbolic power of water – "In the imagination of the twentieth century, water lost both its power to communicate by touch its deep-seated purity and its mystical power to wash off spiritual blemish," Illich writes (Illich 1986: 75–76).

What, then, of the pool? Is it really still possible to claim a link here to water's mystical or mythical properties? Or is immersion in pool water somehow, sterile, separate? Man's ever-increasing domination of water manifests at the pool in various ways, after all: in the containment of the water, in the regulation of its temperature, in its adulteration with purifying agents such as chlorine. But its power as a symbol – in art, literature, the imagination – derives at least in part from its relationship to water itself. The pool is irrevocably linked to water and its metaphors and meanings. Swimming laps is not a form of baptism, not even as similar to the act as certain kinds of wild immersion may be – but it is still a form of immersion. As Thomas Farber writes: "Though Ivan Illich may be right, though humans have transformed water into a chemical detergent [...] still it is held to possess spiritual powers" (Farber 1994: 23).

Reading this reminds me of Mary, an amateur triathlete and an employee of the Church of England, who religiously swims the same distance – 2,000 metres – whenever she goes to the pool. When I ask her what she thinks about for the duration of those 2,000 metres, she says that sometimes she's working out solutions to problems while she's swimming laps, but

> other times, I just pray. I think there is a spirituality about water. I'm coming from a Christian perspective, but I know other people, Christian and non-Christian, that would say there's definitely a spirituality to water. And of course the whole symbolism of baptism... there's definitely something spiritual about it
>
> (Mary, interview, July 2014)

v

Whether you believe that immersion signifies the fulfilment of some prenatal fantasy, or that it has some healing power, or that there is something innately spiritual about it, there is a strong sense that water can convey a message, and that, in fact, to swim may be to embark upon a journey not just up and down the pool, or across the lake, but along the watery lines of that message. As Laing writes:

> There have been times when, sunk in a river or a chalky sea, I have felt the past rise up upon me like a wave. The water has loosened something [...] The present is obliterated, but what the eye sees, what the ear hears, is not possible to share
>
> (Laing 2011: 113)

There are times, in other words, when to swim is to reach or be carried back, or out – out of time, out of body.

This reaching out or back is not always so dramatic as the *nostalgie de la boue*, the unconscious desire for a return to amniotic waters, the perception of a wider environmental past. Sometimes the past that the water has loosened is a more recent and palpable one: a childhood memory, a moment or feeling from the swimmer's bank of retrievable recollections. The connection between water and childhood, as opposed to some more unknowable pre-childhood existence, is simpler to grasp.

When I ask people about the origins of their relationship with swimming they often recall that they learned as children, even if they didn't develop a regular swimming practice until later in life. Hearing these stories, I read parallels into my own. My father was a swimmer in high school, recalls long hard practices after school, 10,000 metres up and down the pool. My mother, on the other hand, doesn't know how to swim: she never learned and then eventually, she says, it just felt like it was too late. But it was not too late for me. I had swimming lessons, which my mother ferried me to, anxious and eager in equal measure to see me develop a fluency in the water. My memories of these lessons are physical, not technical: the chlorine stinging my eyes, the pervasiveness of the water, the way it always found its way up my nose, the feeling of alarm tinged with thrill at finding myself in the deep end, the discomfort of too-tight goggles. I remember being particularly afraid of an instructor called Char, who had a booming, husky voice that conveyed disappointment or disapproval even if it was only saying "good morning!" She wore a clear plastic visor and had rough, saggy brown skin from too many years of sunbathing, too many years of pacing the deck, barking at frightened children, compelling them into the water, willing them to adopt better, more fluid form.

I must have absorbed some of her instruction, but even so I didn't swim much when I was a child, though I was very often in the water, splashing around, boogie boarding, playing games with friends. I had no urge, nor did I receive any encouragement, to join a swim team or formalize the relationship in any other such way. Swimming, as far as I was concerned, was synonymous with play. It was the thing I did for fun when friends had birthday parties; it was sitting in the hot sun until I couldn't bear it and then leaping joyfully into the cold water, practising dives, feeling the disorienting whirl as, after a particularly flamboyant swan-dive, my body flipped underwater, and I opened my eyes in a cave of bubbles, and I didn't know which way was up or down ("to swoon, to lose consciousness...").

Observing lap swimmers, it is not immediately obvious that they treat the pool particularly as a place of playfulness, even if they once did. Watch them go up and down, up and down; watch them scowl at each other when one interferes, even inadvertently, with the planned trajectory of another, or when some environmental factor – water temperature, air temperature – deviates from the norm. Watch them sweat, wheeze, stretch, spit, check the clock, haul themselves inelegantly from the water and onto the deck, where their full adult weight becomes again apparent to them. Watch them follow predictable paths and flows: they have learned the structure of this space through movement and practice, they know and mostly adhere to its rules. And yet even so, swimming may be an expansive or evocative experience: even at the pool, to swim can arouse the memory of some childish playfulness, or, further back, some "pre-literate world". And it's not only association or metaphor that causes this reaching back and out, but also the sensory experience of being in the water: what it looks like, what it feels like, what it smells like.

In *Topophilia,* the geographer Yi-Fu Tuan explores the relationship between the senses and the perception of the environment. "Perception is an activity," he writes, "a reaching out to the world" (1974: 12); odour, for instance, "has the power to evoke vivid, emotionally-charged memories of past events and scenes" (1974: 10). When I read this I'm put immediately in mind of chlorine. Chlorine is representative of our desire to purify water, and yet our inability to do so completely, not only because it cannot prevent certain intrusions (a ball of snot, a plaster, a hair, a piece of paper) but also because it is itself an intrusion, a pollutant. It can assault the senses, aggravate them; I met a swimmer recently who has had to give up swimming in pools entirely after developing an allergy to chlorine. And yet it can also evoke memories and feelings far beyond any immediate discomfort – indeed, it can, through its evocation of these feelings, actually be itself a comfort.

Which makes me think of Hank, freshly arrived in the city. He speaks to me about identifying a soothing sense of familiarity at his new pool, rooted in childhood and evoked largely by the smell of chlorine:

> The changing rooms of these pools always tend to be the same. [...] And there's that certain smell to them as well – and when you're getting changed in there, it makes you feel like you're a kid. [...] And that makes it very comfortable.
>
> (Hank, interview, July 2014)

vi

Sometimes, a few hours after my morning swim, I'll rest my chin in my hands, and I will catch a whiff of something. It's chlorine, mixed with, marked by, time; chlorine diluted by a few hours since my last immersion, diluted by the smell of the hand soap in the bathroom, the shampoo I used in my hair after my swim, the garlic I sliced at lunch. An imprecise mixture, an inconsistent potency, but the thing that's consistent is the *tone* of the chlorine, which is muted and therefore somehow more evocative than if it was fresh.

It reminds me, first and most viscerally, of the pleasure of a swim, what it feels like to be in water, the comfort of the environment; if I happen to be somewhere strange or uncomfortable (a dentist's waiting room, a cramped bus) the smell is soothing, a reminder of a familiar place, a familiar state. At the same time it reminds me of my childhood, and not only of the happy carefree hours spent lolling around cold turquoise pools on hot bright California summer days: it reminds me too of Char's gravelly disapproving voice, those lessons during which I grew aware of myself, during which I was afraid, not of drowning, not of the deep end, but of my own bodily limitations. It reminds me of the discomfort of finding myself in an environment in which breath could not be drawn at will but had to be planned and executed precisely, an environment in which the effort I put in did not seem to match the results I got, the speed I attained. I swallowed water, came up spluttering, clung to the side of the pool and yearned to be back on land. In the car, on the way home, I would be heavy and hungry and sleepy. Sunburn, the sting of chlorine in my eyes, the feel of it at the back of my throat. This was the taste of failure, or so I thought.

So I put my hands to my face: I smell pleasures, anxieties. Then, too, there is the gladness of having a history at all, of having an ongoing relationship with water, of both knowing it – even though it cannot be fully known – and feeling, in a strange sense, that it knows me – even though this is impossible.

Sprawson, quoting Paul Valéry:

> my sole pastime, my only sport, was the purest of all: swimming. It seems to me that I discover and recognise myself when I return to this universal element. My body becomes the direct instrument of my mind, the author of its ideas. To plunge into water, to move one's whole body, from head to toe, in its wild and graceful beauty, to twist about in its pure depths, this is for me a delight only comparable to love.
>
> (Sprawson 2013: 101)

vii

"When I was twelve," writes the artist and ex-competitive swimmer Leanne Shapton,

> a coach remarked I had a "feel" for the water. After basking in the attention for a moment, I understood exactly what he meant. I still do. It's a knowledge of watery space, being able to sense exactly where my body is and what it's affecting, an animal empathy for contact with an element – the springing shudder a cat makes when you touch its back. When I'm dry I bump into things, stub toes, miss stairs. I prefer the horizontal, feet up, legs folded over armrests, head propped sideways on my elbow. I don't understand how to really draw until a teacher says, "Imagine you are running your hand over the surface of what you are drawing".
>
> Shapton (2012: 210)

viii

From fieldnotes, Friday 25 April 2014

> H. joined my lane after I'd had it to myself for awhile; at one point we were both resting at the wall and she said, "the water feels like treacle today!" It did for me too. Treacle is a good description – like to get through it at all requires great effort. That's the thing with the consistency or feel of the water – it sometimes seems to fight you the whole way, while other times it seems to give way, to even actively propel you.

*

When it's good it's great. To know the water as friend, not foe; to feel your way through it, to work with it, to carve something beautiful – *a swim!* – out of it: this is what makes any drudgery, any discomfort, worth it. But water is multifarious, relative: on some days silk, others sludge. To swim in open water perhaps makes this harder to see, because of course in an ocean or a river which is in constant movement the water's character is constantly, visibly, changing, but in a pool, where the water is carefully regulated, purified, kept at a consistent temperature, it becomes suddenly clear how much of a body's perception of water has to do with the relationship between body and water and environment on any given day; how much, in other words, it is unfixed, fluid, affected.

Even, or perhaps especially, the most fluent swimmers rely upon an understanding that this relationship is variable. In the opening to his autobiography, the Olympic swimmer Ian Thorpe describes a process of feeling his way through the water, feeling the water's way around him:

> When I first dive into the pool I try to work out how the water wants to hold me. If I let it the water will naturally guide me into a position: a place for my body to settle, resting with my head down almost meditating. [...]

> I don't need someone to tell me that my stroke looks great or that it looks terrible because I have an inner sense of the water and the environment is already communicating with me.
>
> (Thorpe and Wainright 2012: xi–xii).

ix

"To correspond with the world," Tim Ingold has written, "is not to describe it, or to represent it, but to *answer to it*" (Ingold 2013: 108). This is the *respond* of correspondence: a kind of listening, incorporating, of recognizing the world in us, asking a question, waiting for a reply. *I have an inner sense of the water and the environment is already communicating with me.*

Bodies and water respond to each other in various ways: the water accommodating or nourishing the body, the body energised or enervated or in some other way affected by the water. In essence, then, to swim is a form of correspondence – between body and world, self and environment, body and body. The water, felt, touched, touching, facilitates this correspondence, charging it with meaning.

x

What is it to swim?

That question, always nagging at the edge of my mind. A form of correspondence, yes. But, pedantically: if I wade out in a river, hip-deep, and dip briefly below the surface – is that to swim? *No. Yes. No. Yes.*

It's easy to get bogged down in all this, to be carried away on a current. "Water," writes Laing, "is sly" (Laing 2011: 148), and it has slipped, slyly, into my consciousness. What I miss, sitting in that cottage in Wales, watching the rain dribble down the steamy windowpane and the blurred figures of two cyclists climbing the hill behind the house, is largely my routine, and the routine is largely a product of the pool itself. I miss the familiarity of my route to the pool, the blow of hot air as the doors open, the comfort of overhearing changing room conversations about gardening and kids growing up and holidays in Cornwall, the physical challenge, the upkeep of fitness, the constant comparison of myself to myself – *last week I swam that distance faster; but then again, last year I swam it slower.*

But there's also that element of swimming which is less to do with the pool environment specifically and more to do with whatever it is that also makes a quick dip in a frigid, shallow stream a fulfilling experience – I miss that, too.

To swim, then, is to enter a certain state of mind, to take yourself to a certain place, or be taken there – to be both freed and contained by the water itself. The pool and the pond give us different structures and symbolisms, but really they are both sites for the entering of this state of mind – portals, gateways. The pool, in all its humdrum glory, can be just as much a thrilling site for the exploration of the human relationship to water as a river, a cove; indeed, to those who perceive it, in contrast to the wilderness of an ocean or a lake, as a site of "comfort, safety, reliability and calmness" (Foley 2015: 223), the pool may be the only place where this communion – this conversation, this correspondence – with water can take place.

Bibliography

Cheever, J. (2009), "The Swimmer", in Cheever J. (ed.), *John Cheever: Collected Stories & Other Writings*. New York: The Library of America, pp. 726–737.

Deakin, R. (2000) *Waterlog*. London: Vintage.

Eliade, M. (1996) *Patterns in Comparative Religion*. Lincoln: University of Nebraska Press.

Farber, T. (1994) *On Water*. Hopewell, NJ: The Ecco Press.

Ferenczi, S. (1989) *Thalassa: A Theory of Genitality*. London: Karnac (Books) Ltd.

Foley, R. (2015), "Swimming in Ireland: Immersions in Therapeutic Blue Space", *Health & Place* 35, pp. 218–225.

Freud, S. (1997) *The Interpretation of Dreams*. Ware: Wordsworth Editions.

Illich, I. (1986) *H2O and the Waters of Forgetfulness*. London: Marion Boyars.

Ingold, T. (2013), *Making: Anthropology, Archeology, Art and Architecture*. Abingdon: Routledge.

Laing, O. (2011) *To the River*. Edinburgh: Canongate.

Murdoch, I. (1993) "'Taking the Plunge': Review of *Haunts of the Black Masseur* by Charles Sprawson", *The New York Review of Books* 40: 5 (4 March) [online]. Available at: http://www.nybooks.com/articles/1993/03/04/taking-the-plunge/ [accessed 19 February 2016].

Shapton, L. (2012) *Swimming Studies*. London: Penguin.

Sprawson, C. (2013) *Haunts of the Black Masseur: The Swimmer as Hero*. London: Vintage.

Strang, V. (2005) "Common Senses: Water, Sensory Experience and the Generation of Meaning", *Journal of Material Culture* 10: 1, pp. 92–120.

Thorpe, I. and Wainright, R. (2012), *This Is Me: The Autobiography*. London: Simon & Schuster.

Tuan, Y. (1974) *Topophilia: A Study of Environmental Perception, Attitudes, and Values*. Upper Saddle River: Prentice-Hall.

Section four

Health and wellbeing

13 Picturing dis/comforting geographies

Place, *punctum* and photography

Andrew Gorman-Murray

This chapter addresses dis/comforting geographies via an exploration of my own practices of engaging with place through therapeutic photography. I seek to position this piece at the junction of cultural geography and therapeutic photography (which I will explain below), employing a practice-based approach to understanding dis/comforting geographies. This chapter largely comprises a photographic essay, offering imagery that materialises my dis/comforting encounters with local environments in the form of visual language. I choose these words quite deliberately. The images are not just representations of my affective relationship with, detachments from or attachments to place(s). Rather, I argue that these photographs, produced through processes of self-directed therapy, are the very objects that embody and crystallise how I sought to work through my personal geographies of comfort and discomfort – my sense of self, place and wellbeing together – at a particular 'fateful moment' of life. The psychologist Judith Weiser describes such photographs as 'representational objects', using them in her work with clients (Weiser 2014: 165). In this textual introduction, I provide context for the photographic essay that follows. I delineate an understanding of dis/comfort, self and place; outline the circumstances that led to my engagement with place-based therapeutic photography; describe the type of therapy employed (existential therapy with a phenomenological approach); explain the nature of place-based therapeutic photography and thread it with concepts from photography theory (especially *punctum*); and then give the chapter over to the visual language of the images themselves.

Comfort, as an idea, offers a range of meanings within personal and social registers. It may refer to maintaining a privileged, or at least pleasant, lifestyle: living in 'the comfort zone'. At the same time, the drive or desire for comfort, which is often advanced in popular discourse, also points to its tenuous and liminal qualities: comfort is a set of practices that must be actively worked on to achieve consolation from difficulty (material, social or emotional) and sustain physical and mental wellbeing (Wiles et al. 2009; Bissell 2010). As is demonstrated elsewhere in this volume as well as wider scholarship, the practice and experience of comfort is inflected by social positions such as gender, sexuality, ethnicity, age and disability (Noble 2002; Brickell 2012; Gorman-Murray 2013). If we accept these frames of reference, then comfort is indeed both personal and social: intrapersonal wellbeing entwines our sense of self and our location in social landscapes (Noble 2005). Moreover, comfort is contested by its antonym, discomfort, which signifies unease and anxiety. And yet the ideas and feelings associated with comfort and discomfort need not be mutually exclusive: perceptions of discomfort can heighten understanding of the privileges and benefits of comfort (Holliday 1999; Racz 2015). Thinking about these ideas through a geographical imagination, comfort and discomfort are about the sense of 'fit' between self and place (Gorman-Murray 2009; Butcher 2010). This operates at various spaces and scales: the body, home, city, nation and 'world' (Noble 2002). The chapter takes these ideas forward by reflecting on my own dis/comforting geographies.

I need to offer some personal background so the reader can understand the scope of the chapter: I have bipolar mood disorder, a mental illness typified by severe emotional lability between episodes of mania and hypomania (extreme mental agitation, with or without euphoria), depression and mixed-mood states. Bipolar is a leading cause of disability globally, with individuals encountering significant life adversity (Niztburg et al. 2016), including disruptions to everyday activities, employability, interpersonal relationships, health and quality of life (Bonin et al. 2012). Bipolar has a significant affect on the capacity of individuals to both understand themselves and relate to the world around them (Inder et al. 2008; Chouinard 2012). Vacillating through cycles of mania, hypomania, depression and mixed states transforms one's experiences and perceptions of mind, body, place, time, movement, rhythm and, indeed, reality. Yet, as Vera Chouinard (2012: 144) argues, there is a dearth of work on 'geographies of life with bipolar' (but see England 2016 for an important account of living with bipolar in the geography academy). Individuals may experience both their sense of self and sense of place in radically different ways during different episodes (and due to the effects of mood-stabilisers, e.g. Lithium) (Inder et al. 2008, 2011; Lobban et al. 2012). Consequently, bipolarity has a significant impact on the dis/comfort of place, blurring the experiential boundary between self and environment (Chouinard 2012).

The past few years, since 2013, have been a period of personal upheaval, exacerbating the lability of my bipolarity. This has generated anxiety and compromised ontological security – that is, the circumstances and my response to them has destabilised my sense of security in my-self and my place in the world, my perceived continuity of life events and experiences (Giddens 1991). In *The Divided Self: An Existential Study in Sanity and Madness*, anti-psychiatrist R.D. Laing (1960) called this condition 'ontological insecurity', or a deeply felt personal unease both produced by and producing a fragmentation of one's (register of) experiences and sense of being. For me, intrapersonal conflict has led to further psychotherapy. The therapeutic regime has included counselling and pharmacotherapy, but what is of interest here is the use of therapeutic photography in an effort to improve my ontological security, or my sense of being-in-the-world. The term being-in-the-world will be familiar to existentialists and phenomenologists, and indicates the school of psychotherapy with which I have been working: existential therapy. Rather than foregrounding a medical model of mental illness, existential therapy is a philosophy-based, non-pathologising therapy (May 1983). The therapeutic approach is drawn from existential and phenomenological ideas, and engages the client in a thorough examination of how they understand their existence – their 'being' and their relationship to the phenomena of the world – in order to better consider their choices and values in the effort to create a 'meaningful' life (Spinelli 2015).

Being-in-the-world – a critical concept that has also influenced cultural geography since the development of the subfield of humanistic geography in the 1970s (Tuan 1977; Buttimer and Seamon 1980) – is thus foundational to this therapy, which utilises a phenomenological approach. Deriving from the philosophy of Martin Heidegger (1962 [1927]), being-in-the-world provokes a reconsideration of what it means to 'be' human, particularly the relationality of 'being':

> Being-in-the-world is a term that makes it clear that we are always situated within the world, involved in a meaningful context consisting of people, ideas, places, objects and events – inter-connected and inter-dependent.
>
> (Iacovou and Weixel-Dixon 2015: 17)

The extended term, being-in-the-world-with-others, emphasises the role of relationships in the nature of existence and the ongoing construction of self. The phenomenological method seeks to access and examine our understanding of being-in-the-world through engaging with things as they appear – or more specifically, examining 'phenomena as they present themselves in consciousness as immediate experiences' (Betenskey 2011: 121). Through returning to the experiential context (rather than distanced objectivity), the goal is to open a reflective, therapeutic space in which to consider the significance of personal perspective and interpretation in creating lived experience and the subsequent meaning-making that flows from our being-in-the-world (Iacovou and Weixel-Dixon 2015).

For me, therapeutic photography has played an important role in this phenomenological approach, particularly the self-directed elements of my therapy. Mental health counsellor and art therapist Kathy Malchiodi (2012: 33) defines therapeutic photography as 'the personal use of photographs for self-discovery … for self-awareness, creative expression and self-directed therapy and wellness'. Therapeutic photography is a self-directed practice that does not take place with a counsellor but may be incorporated into discussions within psychotherapeutic sessions (as mine were). It is distinguished from PhotoTherapy and Art-Photo-Therapy, which directly use photographs in psychotherapeutic sessions – to create or explore a visual language to examine memory and unconscious – under the guidance of the counsellor (Weiser 2004, 2014). It is also distinguished from art practice itself – I want to point this out because I am also a practising artist, sometimes undertaking my art practice in an interdisciplinary collaboration with my geographical work (Gorman-Murray 2014, 2015, 2016, 2018). For these reasons it is important to understand that therapeutic photography is self-directed therapy, not art practice with aesthetic directions and intentions.

While the goal of this therapy was to examine my being-in-the-world – more specifically, my lived experience of bipolar – and work towards ontological security and a 'stabilised' a sense of self, meaning and experience, this did not (and does not) necessarily equate to finding or making comforting geographies. Psychotherapeutic processes any of any kind can be discomforting, but existential and phenomenological approaches have been said to be especially challenging for clients. Susan Iacovou and Karen Weixel-Dixon (2015: 10) assert:

> Existential therapy is not a flavour of therapy that everyone finds palatable. It demands much of clients, who must be prepared to wrestle with, and ultimately come to accept, the dilemmas and paradoxes of human existence, to take responsibility for the choices they make (and the consequences that result from them) and to confront absurdity, meaninglessness and the finitude of their own existence with courage and tenacity.

In other words, existential therapy and the phenomenological approach are equally about discomfort and comfort, and the therapeutic process arguably works in and through the tension of dis/comfort. Indeed, existential therapists such as Rollo May (1983) and Victor Frankl (1980) argue that psychological growth requires us to accept all forms of lived experience equally, even the uncomfortable and unpleasant, in order to attain a sense of ontological security.

My examination of being-in-the-world through the practice of therapeutic photography was, and is, no different then – it involves meeting with and working through the tension of dis/comfort to develop a more stabilised (note, not necessarily comfortable) sense of 'fit' between self and place in the context of bipolar. Therapeutic photography, then, is not a contrivance to create banal images of comfort; it is a mode of self-directed therapy that engages with a spectrum of experiences of being-in-the-world, the discomforting, awkward and uncanny as well as the familiar and easy. The images collected in this photo essay are meant to generate a personal yet critical take on the role of the visual in dis/comforting geographies. Comfort and discomfort, like all placed experiences, are spatially and temporally contingent and mutable. For me, picturing a place via therapeutic photography – 'being there', being-in-the-world – meant capturing unexpected and unfamiliar moments in an otherwise familiar landscape. The 'familiar landscape' was the neighbourhood and surrounds in which I live – and have lived for over 15 years – in Sydney's inner western suburbs. In trying to examine my being-in-the-world and my ontological in/securities using therapeutic photography, I sought, perhaps unconsciously, to document at least two kinds of images. On the one hand, uncanny moments – unexpected, sometimes uncomfortable, phenomena in the landscape – and on the other hand, those sometimes taken-for-granted phenomena, which are therefore unfamiliar because they are overlooked in the business of everyday life.

I want to offer the reader a sense of my approach to practising therapeutic photography; I also want to note that these are not 'rules', but rather some of the ways I try to 'get in the mood' to find, take and reflect upon photographs for therapeutic purposes. My practice is about being-in-the-world, so my approach to therapeutic photography begins with walking encounters in my local environment. I take inspiration from the strategy of the dérive, proffered by Guy Debord (1956) and the Situationist International. The dérive is a walking practice that contests the formal (capitalist) structure of the city by engaging in unplanned journeys by foot through the urban landscape and allowing one's path and attention to be drawn by the attractions, encounters and affects therein; in a sense, going where the mood takes you. I use my digital SLR camera for therapeutic photography. This lens-based practice (rather than my iPhone camera) necessitates a particular embodied framing of the world, one where the 'eye' and the 'body' must closely align with the phenomenal world to make images. Taking photographs on dérives, and subsequently (re)viewing and organising them afterwards, involves both deductive and inductive approaches. On the one hand, I deductively work through existential and geographical concepts that inform my observations, such as being-in-the-world, other-than-human-dwelling, the uncanny, freedom and mortality (akin to existential 'throwness'). On the other hand, I work inductively too, and sort any repetitive images from dérives into emergent themes, such as connectivity, grounding, structure, seasons and perspectives. These deductive and inductive patterns are reflected in the representative photographic archive that follows. I keep a thematically organised digital archive, but print particularly poignant images, or whole series, for more intense self-reflection and for discussion in therapy.

Indeed, if the phenomenological method in existential therapy encourages us to engage with things as they appear to us in the world, we also find some helpful insights from writing about photography theory and practice that help us to further interpret (or at least give context to) the phenomena appearing in photographic images. In particular, Roland Barthes' (2000 [1980]) discussion of *studium* and *punctum* is useful for understanding dis/comforting geographies in photographic imagery (see also Rose 2010). I am not suggesting that *studium/punctum* maps congruently on to comfort/discomfort by any means, but rather that *studium/punctum* can help us see how dis/comforting geographies might be simultaneous, and that comfort may issue from uncanny experiences. *Studium* is what is anticipated in a photograph, the meaning that might be interpreted as a result of shared social or cultural knowledge amongst a population. It is that:

> which I perceive quite familiarly as a consequence of my knowledge, my culture. ... It is by *studium* that I am interested in so many photographs, whether I receive them as political testimony or enjoy them as good historical scenes: for it is culturally (this connotation is present in *studium*) that I participate in the figures, the faces, gestures, the settings, the actions.
>
> (Barthes 2000: 25–26)

Studium might be said to be comfortable because of its familiarity – it is a common frame of reference shared across society and culture. That does not mean the content of the image is palatable, of course, as in the case of increasingly ubiquitous war photographs of dead bodies, injured people and bombed structures, although the shock of content may wane with familiarity too.

It might be said, then, that *studium* is what is socially meaningful in a photographic image, but not that which is necessarily personally meaningful – that is denoted as *punctum*. *Punctum* is the element that disturbs, breaks or punctuates the anticipated *studium* of the photograph. In defining *punctum*, Barthes (2000: 26) gives agency to the image itself: 'it is not I who seek it out ... it is this element that arises from the scene, shoots out of it like an arrow, and pierces me'. *Punctum* is described in terms that are unnerving: wound, prick, sting, cut: 'A photograph's *punctum* is that accident which pricks me (but also bruises me, is poignant to me)' (Barthes 2000: 27). *Punctum* is unfamiliar (it even emerges from the scene in the photograph, not from the viewing self) and uncomfortable (it wounds and bruises) and yet is also the element that creates personal meaning in the scene (it is poignant to me). Here, the familiar and comforting is generated by a discomforting experience.

What I want to suggest is that *punctum* is an active element of therapeutic photography, including my own practices. What drew me to look again at familiar landscapes through a lens-based practice was precisely the capacity for *punctum* to 'shoot out' of the photograph and make me look again at the scene. In looking again at the scene, drawn by the *punctum*, I could also begin to look beyond the scene (within the image) to the setting itself (where the photograph was taken) and consider my engagement with both the scene and setting of the photograph (since I was the image-maker). The *punctum* was the dimension of therapeutic photography that enabled a visually-charged, reflective consideration of my being-in-the-world and my lived experience of bipolar. Those unexpected moments paradoxically enabled personal meaning-making in a context of dis/comforting geographies. Importantly, in the case of therapeutic photography – a self-directed practice – meaning is made during both processes of *taking* photographs and *viewing* them again. Recognising and responding to the *punctum* in the photographs allowed me to reflect on my compositional choices – the content, the angle, the framing, why – and thus access my thoughts and recollections of being-in-the-world-*at-that-moment* in order to mediate and stabilise the emotional and experiential lability of living with bipolar. This process of visualising, creating, viewing and reflecting – connected through the arrow's flight of the *punctum* – went some way towards enabling a sense of ontological security and comfort-in-place.

What follows is a visual archive of my practices of therapeutic photography. In light of the above discussion, it is important for me to point out that readers/viewers are not likely to experience the sense of *punctum* that I did with many of these images. *Punctum* is linked to a personal response to an image, and I would not expect anyone else to be pierced the same way. (Hence, it was important to provide the above discussion to contextualise explicitly the contribution of this chapter to dis/comforting geographies.) Nevertheless, given its personal nature, some of you may indeed experience your own *punctum* in and through some of these scenes. In viewing them, they may reveal personal meanings to you. Or they might not; they may only offer the possibility of *studium* for you. So at the same time, to provide some kind of archival frame around my interpreted images, I organised the photo essay through a thematic typology – a typology of the kinds of setting and scenes that pierced me during the process of creating and viewing therapeutic photographs. The thematic titles therefore reflect personal meanings for me and may differ from your own reading of the images. That is fine; reinterpreting the relationship between the photographs and the thematic titles will be part of your own meaning-making. Ultimately, my aim in this chapter is to contribute to our understanding of the role of the visual – as a personal practice – in creating dis/comforting geographies.

Uncanny

Mortality

Freedom

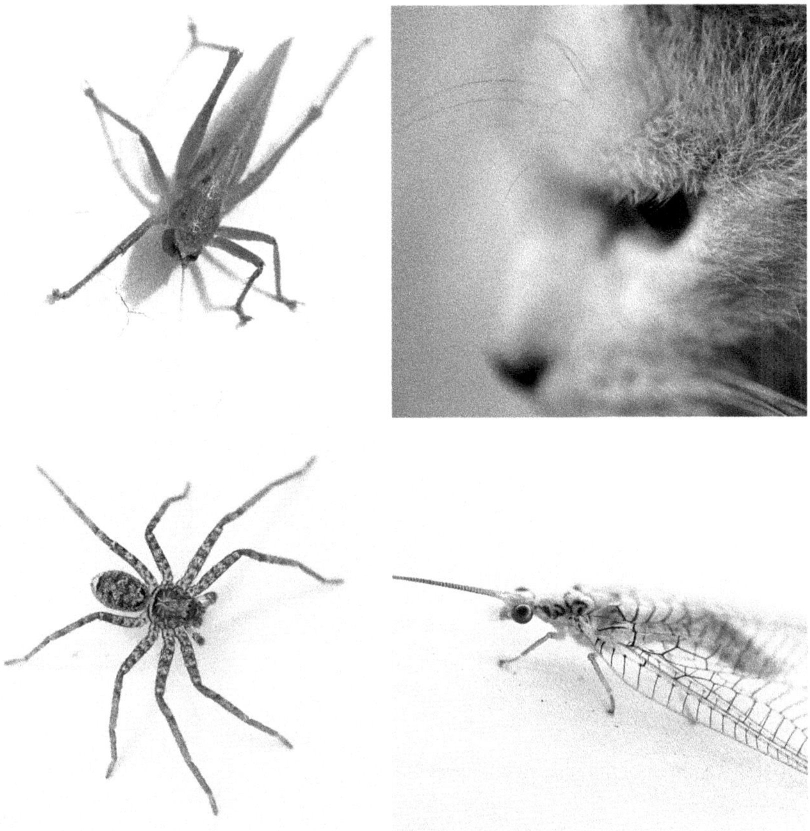

Being-in-the-world-with-others

Perspectives

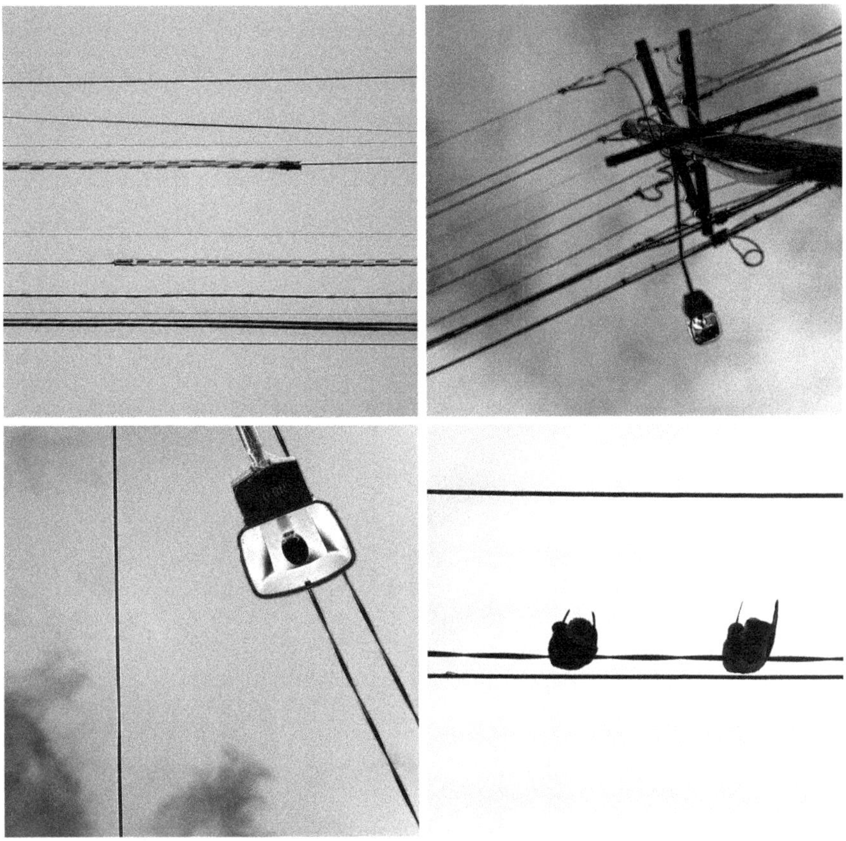

Connectivity

Grounding

Structure

Other-than-human-dwelling

Seasons

References

Barthes, R. (2000 [1980]) *Camera Lucida*, New York, Vintage.

Betenskey, M. (2011) 'Phenomenological art therapy', in Rubin, J. (ed.) *Approaches to Art Therapy: Theory and Technique*, New York, Routledge, 121–133.

Bissell, D. (2010) 'Passenger mobilities: affective atmospheres and the sociality of public transport', *Environment and Planning D: Society and Space*, 28, 270–289.

Bonin, C., Sanchez-Moreno, J., Martinez-Aran, A., Sole, B., Reinares, M. et al. (2012) 'Subthreshold symptoms in bipolar disorder: impact on neurocognition, quality of life and disability', *Journal of Affective Disorders*, 13, 650–659.

Brickell, K. (2012) 'Geopolitics of home', *Geography Compass*, 6, 575–588.

Butcher, M. (2010) 'From 'fish out of water' to 'fitting in': the challenge of re-placing home in a mobile world', *Population, Space and Place*, 16, 23–36.

Buttimer, A. and Seamon, D. (1980) *The Human Experience of Place and Space*, London, Croom Helm Publishers.

Chouinard, V. (2012) 'Mapping bipolar worlds: lived geographies of "madness" in autobiographical accounts', *Health and Place*, 18, 144–151.

Debord, G. (1956) 'Theory of the dérive', *Les Lèvres Nues* no. 9 (Paris, November 1956). Reprinted in *Internationale Situationniste* no. 2 (Paris, December 1958), translated by Knabb, K.

England, M. (2016) 'Being open in academia: a personal narrative of mental illness and disclosure', *Canadian Geographer*, 60, 226–231.

Frankl, V. (1980) *Man's Search for Meaning*, New York, Simon and Schuster.

Giddens, A. (1991) *Modernity and Self-identity: Self and Society in the Late Modern Age*, Redwood, CA, Stanford University Press.

Gorman-Murray, A. (2009) 'Intimate mobilities: emotional embodiment and queer migration', *Social and Cultural Geography*, 10, 441–460.

Gorman-Murray, A. (2013) 'Urban homebodies: embodiment, masculinity and domesticity in inner Sydney', *Geographical Research*, 51, 137–144.

Gorman-Murray, A. (2015) 'Material Conditions in the Post-Human City', in *Thresholds*, group exhibition at Ivan Dougherty Gallery, Sydney, 17–20 June 2015, catalogue by Santamaria, J., Lyra and Spark, R. (ISBN 9780987138255).

Gorman-Murray, A. (2016) '*Thrown-togetherness*, 2015: exegesis', in Crabtree, L., Cook, N. and Davison, A. (eds) *Housing and Home Unbound: Intersections in Economics, Environment and Politics in Australia*, London, Routledge, 18–21, 94–97, 168–171, 232–235.

Gorman-Murray, A. (2018) '*Thrown-togetherness*: queering the interior in visual perspectives', in Gorman-Murray, A. and Cook, M. (eds) *Queering the Interior*, London, Bloomsbury, pp. 15–25.

Gorman-Murray, A. with Brickell, C. and de Jong, A. (2014) 'Over the ditch', in *On Islands Creative Collaboration Festival*, Eramboo Artist Environment, Terrey Hills, NSW, 15–30 November, catalogue by Griffith, S. (ISBN 9781634430210).

Heidegger, M. (1962 [1927]) *Being and Time*, translated by Macquarrie, J. and Robinson, E., London, SCM Press.

Holliday, R. (1999) 'The comfort of identity', *Sexualities*, 2, 475–491.

Iacovou, S. and Weixel-Dixon, K. (2015) *Existential Therapy*, London, Routledge.

Inder, M., Crowe, M., Moor, S., Carter, J., Luty, S. and Joyce, P. (2011) "It wouldn't be me if I didn't have bipolar disorder': managing the shift in self-identity with bipolar disorder', *Journal of Nursing and Healthcare of Chronic Illness*, 3, 427–435.

Inder, M., Moor, S., Luty, S., Carter, J. and Joyce, P. (2008) "I actually don't know who I am': the impact of bipolar disorder on the development of the self', *Psychiatry*, 71, 123–133.

Laing, R.D. (1960) *The Divided Self: An Existential Study in Sanity and Madness*, London, Penguin.

Lobban, F., Taylor, K., Murray, C. and Jones, S. (2012) 'Bipolar Disorder is a two-edged sword: a qualitative study to understand the positive edge', *Journal of Affective Disorders*, 141, 204–212.

Malchiodi, K. (2012) 'Art therapy materials, media and methods', in Malchiodi, K. (ed.) *Handbook of Art Therapy*, Guildford Press, New York, 27–41.

May, R. (1983) *The Discovery of Being: Writings in Existential Psychology*, New York, Norton.

Noble, G. (2002) 'Comfortable and relaxed: furnishing the home and nation', *Continuum: Journal of Media and Cultural Studies*, 16, 53–66.

Noble, G. (2005) 'The discomfort of strangers: racism, incivility and ontological security in a relaxed and comfortable nation', *Journal of Intercultural Studies*, 26, 107–120.

Niztburg, G., Russo, M., Cuesta-Diaz, A., Ospina, L., Shanahan, M. et al. (2016) 'Coping strategies and real-world functioning in bipolar disorder', *Journal of Affective Disorders*, 198, 185–188.

Racz, I. (2015) *Art and the Home: Comfort, Alienation and the Everyday*, London, Tauris.

Rose, G. (2010) *Doing Family Photography: The Domestic, the Public and the Politics of Sentiment*, Farnham, Ashgate.

Spinelli, E. (2015) *Practising Existential Therapy: The Relational World*, London, Sage.

Tuan, Y-F. (1977) *Space and Place: The Perspective of Experience*, Minneapolis, University of Minnesota Press.

Weiser, J. (2004) 'PhotoTherapy techniques in counselling and therapy – using ordinary snapshots and photo-interaction to help clients heal their lives', *Canadian Art Therapy Association Journal*, 17, 23–53.

Weiser, J. (2014) 'Establishing a framework for using photos in art therapy (and therapies) practice', *Arteterapia*, 9, 159–190.

Wiles, J., Allen, R., Palmer, A., Hayman, K., Keeling, S. et al. (2009) 'Older people and their social spaces: a study of well-being and attachment to place in Aotearoa New Zealand', *Social Science and Medicine*, 68, 664–671.

14 Between bodies and buildings

The place of comfort within therapeutic spaces

Daryl Martin

Introduction

In this chapter, I will focus on some of the ways in which comfort has been discussed and reflected upon in my research with visitors and staff members of Maggie's, an organisation which provides practical, emotional and social support for people with cancer, their families and friends. Located principally in the UK, but with Centres now open and in development internationally as well, Maggie's provides places to find advice and information about all aspects of living with cancer (including nutrition and welfare), emotional support from cancer specialists and clinical psychologists, and communal areas in which to meet other people with cancer, or quiet spaces for individual reflection. My research has taken place across seven different Centres, and this chapter presents a selection of data from staff interviews and focus groups with visitors and volunteers. Based on this research, I argue that Maggie's Centres are alternative spaces which complement the services offered in their neighbouring hospitals; they act as counter-geographies to the technologically saturated spaces and institutional timetables that their visitors experience more routinely through the stages of a cancer diagnosis and treatment. Maggie's Centres allow different spatial practices, encouraged through a range of objects and furnishings, and a qualitatively different sense of time (for reflection, for sociability). These buildings offer recognition of cancer as an illness experienced by individuals as well as a disease to be treated through clinical protocols, in which the role of comfort is central to how care is approached.

Maggie's Centres are often iconic buildings, designed by some of the most celebrated architectural practices of their day; as such, they have attracted much professional and media attention and have contributed to wider debates about the efficacy of architecture and design in the treatment of illness. However, the elegance of their structures, and the metaphorical allusions inherent in particular designs (see Jencks 2010, p.15), should not distract us from the other qualities that they help to enact in their working lives. Lees has argued that 'if we are to concern ourselves with the inhabitation of architectural space as much as its signification, then we must engage practically and actively with the situated and everyday practices through which built environments are used' (2001, p.56). In this chapter, I report the experiences of those who use these striking buildings routinely and I draw on comfort as a primary theme with which to describe the buildings' emotional potentialities. In so doing, I locate this research within recent work by geographers on the role of affect within understandings of architecture. These are debates which have sought to complicate our assumptions of what architecture comprises, understanding it 'not simply as an accomplishment (or artefact) of human doing, but as an ongoing process of holding together' through the sociological practices buildings allow (Jacobs and Merriman 2011, pp.211–212). In addition to observing the actions of individuals, the material properties and physical objects located in buildings, researchers should give renewed attention to 'how feelings might hold "big things" together' (Lees and Baxter 2011, p.117, drawing on Rose, Degen and Basdas 2010). This is important when presenting the perspective of those who inhabit the buildings we study in their daily lives (Degen, DeSilvey and Rose 2008), but also in acknowledging the agency of architectural form itself, and in understanding the 'push' that buildings provide in shaping the practices of those who use them (Kraftl and Adey 2008, p.219). Doing so helps us to build a more vibrant understanding of architecture, to 'get up close to buildings as occupied performative events' (Jacobs 2006, p.10), and to recognise the atmospherics of place they can afford.

This chapter begins with a brief description of Maggie's, the origins of its work and the integral role of architecture and spatial design to its remit. The narrative account of a cancer diagnosis by its founder prompts comparisons with how medical anthropologists and sociologists have described the impact of technologies and institutional processes upon the embodied experience of medical diagnosis and treatment (Leder 1990; Mol 2002; Blaxter 2009; Riessman 2015). The chapter then opens out its discussion of comfort as an important thematic when thinking about the experience of care. Theoretically, this draws on David Bissell's understanding of comfort as a 'highly complex sensibility' (2007, p.1697) which can be characterised, concurrently, as 'an objective capacity', 'an aesthetic sensibility' and 'a specific affective resonance', where comfort may be enacted through the relations between individual bodies, objects and environments (2007, pp.1700–1701). Bissell's multiple definitions are set into dialogue empirically with interview and focus group data from my research across Maggie's sites, where the importance of comfort within my participants' experience of their Centres is explored. Ultimately, comfort is presented as an achievement of professional practice and spatial design, and an accomplishment through the co-production of these places by their visitors, and the experiences of hope and hopefulness they bring (Anderson 2006).

Maggie's centres, and an architecture of hope

Maggie's offers a programme of non-clinical care, welfare advice, creative therapies, nutritional workshops and psychological support to those with cancer, their families and friends. Support is provided in a series of buildings adjacent to major hospitals that are staffed by cancer support specialists, clinical psychologists, welfare advisors and volunteers. These services are offered on a non-referral basis, making them freely available to anyone who wishes to visit, without appointment. At the time of writing, 20 years since the first Centre opened in 1996, there are 17 established Centres, including the first international Centre located in Hong Kong, with further buildings in various stages of planning. In Maggie's, architecture is viewed to be an active element of the care provided to those who visit the Centres. Maggie's Centres are distinctive buildings designed by globally renowned architectural practices, including those of Norman Foster, Frank Gehry, Zaha Hadid, Rem Koolhaas and Richard Rogers. Their bespoke designs make all Maggie's Centres different from each other, and often vividly so. Independent of the NHS in the UK, Maggie's complements the clinical services provided in their neighbouring hospitals; although sited next to regional oncology centres and hence located in hospital grounds, the buildings act as points of visual differentiation to distinguish the type of support offered within their walls. Centre visitors are *not* referred to as patients; rather, all labels are removed, visitors are known by their first or preferred names, and it is inscribed within the architectural brief that the buildings be designed such that 'people can read themselves differently, as individuals in usually difficult circumstances, not as patients, let alone cancer victims' (Jencks and Heathcote 2010: 219). Maggie's Centres dispense with visible signs of hierarchical status (there are no staff badges, office signage or formal reception areas), with the spatial scale and interior design evoking a homely rather than institutional aesthetic (for example, long corridors are avoided in the design of individual Centres).

Within a spatial footprint of approximately 280 square metres, each building must provide a range of informal and communal spaces that offer a hospitable area for visitors, many of whom use their Centre as a place to wait and chat with others whilst in-between hospital appointments over the course of a day. On arrival at Maggie's Centres, visitors typically walk into a communal kitchen area, prompting an awareness of the familiar artefacts that get crowded out, or remain backstage, in clinical settings. All buildings contain a variety of private rooms wherein confidential discussions take place (about treatment regimes, psychological issues and economic troubles). In addition, each Maggie's Centre incorporates what the architectural theorist and landscape artist Charles Jencks describes as 'gratuitous spaces: the gardens, the architectural gestures, or sculptures, or collections of nick-nacks that exist for themselves' (2010, p.13). Jencks was the husband of the organization's founder, the landscape architect and scholar Maggie Keswick, who commissioned plans for the first Centre in Edinburgh as a response to her own experience of a cancer diagnosis, and as a critique of the spaces and buildings in which her treatment was delivered. Before her death in 1995, she wrote of her experience, offering a description of a mainstream hospital setting that draws out the wider implications of the places in which we routinely receive medical care.

> The NHS is obsessed with cutting waiting time but waiting in itself is not so bad – it's the circumstances in which you have to wait that count. Overhead (sometimes even neon) lighting, interior spaces with no views out and miserable seating against the walls all contribute to extreme mental and physical enervation. Patients who arrive relatively hopeful soon start to wilt.
>
> (Keswick 1995, p.209)

Keswick sketches an architectural atmosphere which will resonate with many if not most who have visited a clinical setting or hospital within the UK. She identifies the interweaving emotional and physical responses to environmental triggers, and the influence of spatial design and material objects on the individual's sense of (dis)comfort. In her account, space and place matter to the experience of disease, not least because of their relation to the temporalities of diagnosis. Keswick's account prompts us to reconsider the prevalent trope of patient choice characterising discourses of contemporary medicine (Mol 2006), through her analysis of clinical spaces in which the overriding feeling may be entrapment; as one focus group member remembered of her own experience, 'you're going into a room that you haven't chosen to go into, it tends to be an enclosed space with a window that looks on to nothing, to be told things you don't want to hear!'. In response, Maggie's commissions architects to design buildings that seek to empower their visitors, through spaces that are hopeful in their affects. The equation of empowerment and hope can be seen as part of a trend in contemporary healthcare discourses that Petersen and Wilkinson urge researchers to be alert to, as they can become implicated in wider trends towards individualised responsibilities for well-being (2015), not least in the field of oncology care (Brown 2015). However, in Maggie's Centres, through their emphasis on the materiality of care and the interactions between bodies and buildings that they afford, I argue that we can glimpse spatial practices that carefully, in their everyday and ordinary affects (Stewart 2007), facilitate a 'taking-place of hope' (Anderson 2006, p.733).

Institutional bodies and absent lives

Maggie Keswick's narrative echoes many of the themes explored by medical anthropologists and sociologists in their writings on medical diagnosis, clinical treatment and the place of the body within these processes. Sociologists have long argued that the practice of diagnosis serves to embed and embody cultural scripts of medical authority at individual and institutional levels (Jutel 2009), in spite of its complexities – as an early contribution to these debates by Blaxter suggested, diagnosis refers to the reproduction of disease categories and also practical processes simultaneously, with these aspects often in tension (1978). To this conceptualisation of 'diagnosis-as-category' and 'diagnosis-as-process', Jutel and Nettleton add a third dimension where they highlight the 'consequences of diagnosis':

> As soon as it is defined (category), or implemented (process), [diagnosis] categorises health realities in tangible ways, determining who has access to what resources, under whose jurisdiction the management of the condition will fall, and what the individual's experience now means in terms of identity and prognosis.
>
> (2011, p.793)

A diagnosis is a social process, which involves disruptions to the lives within wider networks of carers and friends (Olson 2011), as well as a biographically specific event informed by the individual's previous experiences of the medical system within which diagnosis takes place (Schaepe 2011).

Individuals may draw on a variety of narrative strategies to shape their responses to illness (Frank 1995), but often these come at a later stage. For the individual with cancer, the event of diagnosis has been described as a point of *narrative surrender*, with 'the moment when we are thrown into a new discursive field… [playing] an essential part of the process of being captured by dominant constructions (of cancer, of illness)' (Willig 2011, p.903). Reflecting on her own cancer, Riessman notes the importance of her personal chronicle, a journal composed throughout her diagnosis and treatment, and her subsequent writing as attempts to 'bring some order to a ruptured life' (2015, p.1058), in which conventional temporal categories (of life before, during and after diagnosis) were scrambled, if not erased. Mildred Blaxter's last published piece of writing before dying of cancer was an auto-ethnography in which she outlined how the lived knowledge and rhythm of her illness was subverted by institutional timetables, with her body often experienced as an adjunct to the specialist processes driven by medical technologies and the decision-making practices they served to enact (2009). Her paper is an account of many examinations and diagnostic tests, and the translation of their findings into the patient record, such that the patient's voice assumes a secondary place to the documentary rendering of the illness, and the patient's body disappears behind its visual representation. It is worth considering these issues in turn, in order to provide an analytical backdrop to the services and spaces offered by Maggie's, and how these may be understood as geographies of comfort.

In Blaxter's account, the body must learn to adapt to the institutional timetables that underpin the functioning of the contemporary hospital. This resonates with classic ethnographies within medical sociology, such as Roth's observations of the ways in which patient trajectories are defined by temporal benchmarks of progress towards cure, set by clinicians and then internalised and negotiated by patients in their own articulations of recovery

(1963). There are also resonances with Zerubavel's understanding of the socio-temporal rhythms of organisational life in hospitals, where medical events are stabilised by timetables in order to structure the collective efforts of clinical work, and sustain, in effect, patterns of social control (1979). The complexities of multi-disciplinary team working within contemporary medicine necessitate the imposition of order through the institutional timetable, not least because of the use of expensive technologies which medical specialists rely on; but with these come issues of authority over space and the control of process (Blaxter 2009, p.763). The contemporary hospital environment is saturated with technologies and, as Leder argues, even relatively prosaic technologies, such as 'as the stethoscope, the blood test, the X-ray, allow a kind of dissection of the living body, analysing it into its component parts, exposing what life ordinarily conceals' (1992, p.22). This step beyond the immediately visible and audible signs of illness denotes the radiating out of the meaning of illness as individually felt to the representation of disease on the patient record.

For Blaxter, the alienation she experienced as a patient was due not to the technologies she encountered *per se*, but their translation into the patient record which marginalised her own embodied knowledge and, increasingly too, the expertise of medical staff, whose clinical judgement assumed secondary importance to the documenting of disease in the patient record. Following this, the patient record becomes the repository of an imbalance between lay and professional expertise, where the weighting of judgement means that 'the patient is not perceived as the co-producer of the diagnosis: is not seen as the owner of the evidence, and is not thought to have any special knowledge to contribute to its interpretation' (Blaxter 2009, p.774). Similarly, Riessman draws a comparison between her own written accounts and medical documents, with her own narrative accounts acting as 'a counter-narrative to the context-stripping medical record' (2015, p.1062). The patient record becomes involved, more widely, in what Mol expresses as the 'hierarchy between diverging measurements' (2002, p.63), whether the sometimes incongruous test results between medical specialists, or in the distance between what an image tells about disease, and how that telling is understood and its meaning negotiated by the person who is ill.

These accounts throw into sharp relief the tensions between locating the bodily experience of illness within systems of meaning, however open to contestation and negotiation, and maintaining a sense of that experience in non-representational ways. Certainly, they add weight to Mol's arguments for a richer understanding of the ontologies of disease to supplement the epistemologies with which bodies are typically understood in clinical protocols (2002). Mol argues that we should adjust our understanding of disease as something that is done, enacted by institutional processes and environments as well as individual practices, in addition to something that is known, albeit differently, by patients and professionals. To do so entails taking account of the materiality and spatiality of healthcare settings (see also Schillmeier and Domènech 2009; Schillmeier and Heinlein 2009), and understanding that care, like life itself, is a 'matter of chairs and tables, food and air, machines and blood' and the embodied experience of all these things (Mol 2002, p.27). Mol's argument runs alongside and extends those of Leder (1992), who suggests that modern therapeutics are underpinned by understandings of the body as an inanimate machine which needs to be broken down into its component parts, analysed and corrected by intervention. He goes so far as to suggest that this model of the patient has influenced not only formal classifications of illness, but also the spaces in which treatment occurs, and the spatial practices they encourage (Leder 1992, p.33). In a doctor's examination room, the physician takes an active role and the patient is reduced to passivity; the body 'becomes a collection of organs, a mass to be studied and palpated' (Leder 1990, p.98). As well as lending a disempowering dynamic to the encounter for the patient, this also results in a heighted awareness of her body as ill-functioning, provoking 'a bodily thematization' (1990, p.87) or abstracted understanding of bodies which feedbacks and 'can itself bring about dysfunction' (p.85). This dys-appearance is individually felt but also a social process, brought into being, in part at least, through clinical interactions and the medical gaze enacted through processes and technologies such as the X-ray, the blood test, and the routine examination.

However, Leder is clear that, depending on institutional context, the diagnosis can be part of a healing process rather than the generator of fear and confusion (1992, p.29). His critique of modern medicine seeks a working with rather than rejection of it completely:

> *To attend to the lived body is not to forsake the tools and learning that Cartesian medicine has provided. It is merely to refuse to grant this mechanical wisdom the status of a ruling paradigm.* Instead, this wisdom takes on the humbler role of a regional method, elucidating the lived body from one possible perspective, but incapable of capturing its multifaceted richness.
>
> (Leder 1992, p.31, original emphasis)

The configuration of clinical medicine as one approach amongst others in the treatment of disease, in order to do justice to the multiplicity of ways in which that disease is enacted through the multiple selves of the patient, echoes Mol's call for closer attention to the embodied experience of care (2002).

But we can extend this argument, and call for different forms of space in which we recognise the multiplicity of social identities that those with any illness may carry with them. These differential spaces might start with the individual not as patient (as will still be the starting point in their hospital biographies), but rather as someone else; a child, a parent, a friend. And this is where Maggie's plays a part in rounding out the experience of care within a cancer diagnosis and treatment, not least through a spatial form that is multiple in itself, as in Charles Jencks's characterisation of the Centres as hybrid, a kind of non-type: 'like a house which is not a home, a collective hospital which is not an institution, a church which is not religious, and an art gallery which is not a museum' (Jencks 2010, p.14). Maggie's Centres are disarming buildings and spaces, and deliberately so: they have been designed to place the visitor at a very different position to the service, the organisation and the physical environment than he or she will encounter elsewhere in the experience of cancer. They evoke different responses: Maggie's services are hosted in buildings which can be familiar *and* unfamiliar in architectural form to visitors. That is, some Centres are converted buildings, familiar forms that evoke domestic associations; other Centres are iconic buildings, with a sculptural and expressive exterior but with a more homely feel inside, whereas other Centres hold these two qualities, of familiar and unfamiliar spatial forms, in balance (Figures 14.1 and 14.2). Through this dialectic of familiar and unfamiliar form, Maggie's Centres are comforting *and* challenging spaces, often at the same time, as recounted by Centre staff and visitors alike.

Placing comfort

Before describing how comfort was discussed by participants in this research, a brief reflection on the work of David Bissell will help to elaborate how we might take ideas of comfort further in examining their value for understanding the therapeutic potentialities of place. Bissell has characterised comfort as 'a highly complex sensibility' (2008, p.1697), with the body at its nexus and involving the affective intertwining of material objects, physical environments and culturally derived semiotics of comfort. Its deceptively

Figure 14.1 Exterior of Maggie's Gartnavel, Glasgow.

Figure 14.2 Interior image of Maggie's Gartnavel, Glasgow.

simple colloquial connotations carry within them a nested set of meanings, and Bissell identifies three levels from which our everyday understandings may be composed. First, comfort may be recognised as 'an objective capacity', where the state of being comfortable is devolved onto material things: 'comfort in this sense is objectified and made tangible and connoted with an object: as a particular attribute of the object itself' (2008, p.1700). So, one example – and this is the example Bissell uses throughout his article – would be the chair; how comfortable one feels in any particular chair is often taken to be an attribute of that chair itself. Second, comfort may understood as 'an aesthetic sensibility', a definition that captures 'a sensation of being-at-one with the immediate environment that... steps beyond the biomechanical by considering the semiotics of comfort as generative of particular aesthetic meanings' (2008, pp.1700–1701). By this, Bissell notes how an individual feeling of comfort results from intentions of designers to engineer in comfort to a particular object or a specifically designed space; architects are mentioned by Bissell as a prime example of a profession often charged with the creation of comfort through their work. Third, building on and departing from these previous characterisations, comfort may be thought of as 'a specific affective resonance', a relational quality enacted through the interaction of bodies, artefacts and environment. In this understanding, 'comfort as a sensibility is not captured, enclosed, and objectified through specific technologies or objects but can *circulate* between and through both objects and bodies' (2008, p.1701, original emphasis). This offers a sense that comfort is distributed through a variety of human and non-human actants, and plays also on our expectations of the feelings that may come from the assemblage of particular elements and environments. These three understandings interweave within the same comforting objects and spaces, and, as will become apparent, help us to recognise the different registers in which comfort is experienced and anticipated in Maggie's Centres. In the quotes that follow, pseudonyms are used, and identifying information about particular Centres removed to maintain their anonymity.

An objective capacity

Thinking about comfort as an objective capacity, residing in things, was a recurrent theme in the reflections on the Centres by visitors, volunteers and staff. As noted above, Bissell used the example of the chair to explicate the meanings of comfort, and, indeed, chairs and their capacity to provide comfort were repeated points of discussion across Centres. This reflects the discomfort and pain that individuals with certain cancers can feel and the resultant importance of chairs throughout their treatment. Anne Massey, introducing her cultural history of the chair, suggests that chairs both help to structure the dynamics of clinical interactions – whilst the doctor or nurse has the freedom to stand or sit in diagnoses or physical examinations, the patient is usually seated when they are not lying down – and also articulate the physical pain of the ill body. Massey associates the early stages of writing her book with the pain she observed in her partner's bone cancer, the frustrations of trying to find a chair which would make him feel comfortable, and the social implications of that; his 'inability to sit in a chair', she notes, 'reinforced his invalidity as a person, unable to join in accepted social interactions or be a good patient. He couldn't simply *Take a seat*' (2011, p.12, original emphasis). Certainly, within my research across Maggie's Centres, when images of the internal spaces of other Centres were introduced as prompts for discussion in focus groups with visitors and volunteers, chairs were often the first items that participants focused in on and judged in terms of the comfort they might offer.

The kitchen area is typically the first space a visitor sees when arriving at a Maggie's Centre, with a large kitchen table being a visual clue that the Centre is a different type of environment for receiving information and support than found elsewhere in their treatment. By placing such a domestic piece of furniture at the entrance, the intention is to facilitate a feeling of being-at-ease for visitors. The use of smaller objects throughout the Centre serves to reinforce this. The use of books provides an interesting example; in response to my question asking how Maggie's Centres differ from other spaces in which he receives care and support, Tom replied:

> when I go into a GP clinic or urology clinic, I'm in a room in which there are books on shelves I'm never going to be able to read or understand, there are devices around here that I know not what they do, nor would I interpret, be able to interpret what results come from them. And so I feel like I've gone into somebody else's house, that isn't that comfortable

> with my being there. It's OK for the ten minutes you're here, but I want you out. There's a feeling, and there is that remoteness or alienating feeling that these guys know better, and especially when they get onto the topic of telling me what I should do in the light of the books on the shelves and the devices that they've used on me. But that's not here, everything is familiar.

Maggie's Centres have books on shelves too, typically texts on cancer and its psychological and social effects for a general readership. And alongside these will be books not on cancer, but on art, poetry, children's literature and fiction. As opposed to the medical textbooks Tom mentions, these are books that do not exclude visitors, and thus objects being deployed to different ends. In the staff interviews, Nicky, a cancer support specialist, also discussed the place of books in the room we were sitting in for the interview:

> [It's] very rare for anybody to ever pick a book off the shelf in here... but it feels comfortable to sit in here and talk. There's familiarity to the things, the positions we're in when we're talking. It's very different to anything that would exist in a clinical setting or otherwise, you're either at different heights, you're sitting in front of them, there's a desk between you, all those things, this is unusual... that actually you could be sitting pretty much, in very close proximity to somebody, getting into a dialogue even like this.

The proximity in conversation, enabled by the placing of certain objects and pieces of furniture and the avoidance of other objects with closer associations to the hospital setting, enacts a sense of familiarity which is deliberate – artfully done, even. That is, it arises from the professional practice of Centre staff and the attentiveness of volunteers, as well as the material environment in which familiarity has been designed in, allowing visitors to be comfortable there, and to read the space in particular ways. This intentionality of design leads us to the second aspect in Bissell's formulation, that of comfort as an aesthetic sensibility.

An aesthetic sensibility

Understanding comfort as an aesthetic sensibility refers to the ways in which our experiences of space are anticipated by architects and interior (and landscape) designers. Thrift has argued that contemporary environments are notable for the degree to which affect is engineered into the built fabric, shaping the mechanics and textures of daily life (2004). Whilst his scale is that of the city, this translates to the experience of individual buildings. In the case of Maggie's Centres, their spatial designs and material objects play on cultural scripts of domesticity and the feelings of comfort that these can induce. Mandy, a cancer support specialist, talked about the efficacy of designing in familiarity and domesticity within the building and its interiors. She discussed this in the context of her previous experience of working in other settings:

> I thought when I worked in the community, going to visit people at home was you getting them in their comfort space and their comfort zone, getting the immediacy of it, all of that, but actually I get more of that in Maggie's than I did at any other setting.

She continued to explain that:

> I'll have very, very deep conversations with people, very, very quickly. I'll get to that point, either on a first visit, or maybe the second or third time I've spoken to somebody, whereas before, in other jobs, I would have had to have built up that relationship to get to the point of having that conversation. I think the difference is the environment does feel comfortable and homely to people, so they do have a sense of rightness in being here, but I think because they're not in their house, and they're freed up from all of the constraints that that brings. If you're sitting in your own living room surrounded by the photos of your family, or knowing that your husband's in the next room, I think it still does influence and constrain what you're saying and what you feel able to share.

Here we find an argument for the capacity of spatial design and architectural atmospherics to afford effective professional and therapeutic practices. Mandy's comments bring to mind Jencks's statement, quoted earlier, of Maggie's Centres as hybrid buildings which allude to familiar architectural forms without being reducible to these; in this case, 'like a house which is not a home' (Jencks 2010, p.14).

And yet, as Jencks continues, they may also engage with elements of a hospital setting without its institutional feel, a church without its religious connotations and an art gallery without the atmosphere of the museum. The result is a series of buildings that are often unconventional in their external appearance, sculptural and challenging pieces of architecture. They are unfamiliar in form and dissimilar in aesthetic to the buildings they neighbour. Indeed, this idea of a challenging architecture is inscribed in the brief common to all Centres, where it is stated that

> these little buildings should not pat you on the head… They should rise to the occasion, just as you, the person needing help is having to rise to the occasion of one of the most difficult challenges any of us is likely to have to face.
>
> (in Jencks and Heathcote 2010, p.219)

In every focus group and staff interview, I asked participants for their thoughts on the importance of this statement; Mandy replied that:

> in terms of [this city's] styling and architecture and what people of [this city] are used to, this is not a building that necessarily sits comfortably in [this city]. It doesn't sit uncomfortably, but it's not a common building, so it challenges people because it looks completely different to what they're used to being in. Does it need to challenge people? I don't know if I would say it needs to challenge people, I think what's helpful about that challenge is by helping people to get used to that new space, and something that is not as immediate for them, helps them to realise their own adaptability.

Certainly, the balance between an architecture that challenges and an architecture that comforts is weighted differently by different people, and so points to the contingencies of how comfort is felt. The following discussion between two focus group participants in the same building highlights this:

ANDY: I think the first time I saw it I was impressed by the building, but I was also a bit intimidated by it... but I came in, and yeah, it was impressive. I think the lack of colour made me feel a wee bit uncomfortable, because then it was really stark, wasn't it? They've only recently just started to get the pictures on the walls and things like that, it was very clinical then, it's warmer now, but it still needs a bit more work to it, I suppose it's a work in progress, isn't it really?

TOM: No, I like it for its, in that respect, silence. It's like yet to be written upon, it's that bare. You're coming in, you can write your story on these walls because it's waiting for you. I kind of shrink a little from the come and have a cup of tea... this kind of cosiness thing.

Put simply, people respond differently to the same place and feel different levels of comfort in different types of space. We bring with us a history of affective dispositions and embodied tastes crafted by places we have previously known, and that shapes our experience of any space new to us. Despite the best plans by designers, comfort cannot be predetermined and, as Bissell reminds us, is reliant on bodies themselves to facilitate comfort when understood as 'a specific affective resonance' (2008, p.1701).

A specific affective resonance

In this third meaning of comfort, Bissell suggests that we

> move away from comfort as objectified or as intentional and instead present comfort as... more a set of anticipatory affective resonances

> where the body has the capacity to anticipate and fold through and into the physical sensation of the engineered environment.
>
> (2008, p.1701)

Comfort can be seen as socially situated, with due consideration given to spatial form, material artefacts and the lived experiences of the people that circulate and dwell within. This raises questions of comfortable inhabitation of architecture, which is 'not merely a technical or symbolic achievement', but rather a capacity of a building which 'constantly emerges through ongoing, dynamic encounters between buildings; their constituent elements; and spaces', including a range of human and non-human agents (Kraftl and Adey 2008, p.214). Certainly the embodied experience of comfort needs to be emphasised; comfort is experienced in multi-sensual ways, with the body drawing on a range of our senses when being and becoming at ease, and responding to the designed environment (Pallasmaa 2005). Architecture is experienced in a visual register, but we need to remember that 'visuality is always multimodal' (Degen, DeSilvey and Rose 2008, p.1909) and understand the 'more-than-visual' knowledges that architecture allows (Paterson 2011). Maggie's commissions visually striking buildings, but what can we hear, smell and taste in these buildings? How do they *feel*? Or, more precisely, how does the feel *of* a building coalesce around the feelings individuals experience *in* and *about* that building to combine into an affective geography (Rose, Degen and Basdas 2010)?

Nicky, a cancer support specialist, associated the sounds of a Maggie's Centre with comfort: 'the space ... feels so very different, it feels, it's comforting. There's a racket, you hear kids, but there's something very nice about knowing that that noise out there is non-clinical as well'. For Nicky, the non-clinical feel was also helped by the fact that a Maggie's Centre doesn't smell like hospitals, where the same smell of disinfectants from NHS site to NHS site adds to the environmental triggers that disempower patients. Maggie's incorporates nutritional advice, accompanied by practical workshops, and so the Centres are places that visitors associate with food and drink, and the comfort they provide in sociable settings. Building on Paterson's call for an awareness of the 'distinct but interrelated haptic sensibilities' we bring to bear on our relations with place (2011, p.264), we might stress the tactile qualities of comfort. In another Centre, Josephine, a cancer support specialist, described the south-facing aspect of their kitchen as helping visitors, 'because people know they can sit in the kitchen and feel the warm sun on their back and enjoy that feeling of warmth and comfort'.

In speaking about the emotional qualities of her Maggie's Centre, one focus group member, Marie, introduced me to a Welsh word, 'cwtch', which she defined as 'like a cuddle... cuddle and comfort, cuddle and comfort'. In separate and very different buildings, the idea of an embrace was used to describe the emotional warmth that visitors receive from their

Centre: Anne described her Centre as 'like a warm hug, you just come in here and [are] sort of enveloped in something, like a warmth, magic, warm feeling', whereas in a different site Nicola suggested the building itself was 'comforting, like a hug, when you get your cancer diagnosis'. Rather like the Danish word 'hygge', which Bille has characterised in terms of feelings 'of sheltered-ness, nested-ness and well-being' (2015, p.59), cwtch is an affectionate word which connotes security and warmth, and can also be important to the building of a sense of community. Further discussion amongst Marie's group suggested the word 'cwtch' as applying equally to the comforting feeling one gets with an embrace of a loved one, but also with the experience of certain loved places, akin to the intensities of topophilia (Tuan 1974). Its double meaning captures the interplay between people and place, interpersonal interactions and physical environments, bodies and buildings, which marks the emotional qualities of Maggie's Centres for visitors.

These are buildings that are affectively charged for visitors, and yet it should not be forgotten that this emotional attachment is something which is collectively achieved. Comparing the experience of working in two different Centres, with different spatial layouts and architectural aesthetics, Nicky suggested that:

> people do seem to get very comfortable in that building [another centre], very much quicker, they don't need a guide towards their comfort, whereas in this building they do, and maybe there's something in that, because that ... they're seeing us as a navigator... They're engaging with us and they're identifying who can help them with that.

Maggie's Centres can exude an ambient sense of ease within their walls that is gracefully achieved, but which involves direction, negotiation and labour. The kitchen table, so important in terms of dispelling the institutional expectations Centre users may carry with them on a first visit, is a case in point. The kitchen table is both a communal place of relaxation for visitors and a site of difficult and emotionally demanding work for staff and volunteers; the comfort visitors may feel here is managed, with a performative element or kind of choreography (Martin 2016). It is here that we glimpse examples of what Riessman terms the 'micro-work of interaction' so crucial in her own experience of cancer; for her, the routine emotional labour of radiation nurses and occupational therapists, alongside conversations with fellow patients, 'performed the familiar sociability of everyday life' and established feelings of comfort that affected her deeply in her hospital stay (2015, p.1060). Riessman's account, and my observation of many encounters in Maggie's Centres, brings us to an awareness of the potentialities of comfort in therapeutic encounters, and opens up an understanding of comfort as an ongoing achievement, as a collective accomplishment (cf. case studies in Kraftl and Adey [2008]).

Conclusion: hopefulness within, and through, place

As stated at the outset of this chapter, Maggie's offers practical, emotional and social support for people with cancer, their families and friends, and values architecture and interior design to an untypically large degree as means of achieving this. In conclusion, I turn to a carer's perspective: David, who visits the Centre to accompany his partner, suggested that:

> from a carers' point of view you want somewhere you feel, sit down, feel comfortable, if I want to go and have a chat with... one of the people who are helping, they're there, they know, but they're not in your face. And they've always got a cup of tea, there's a biscuit. You just feel like you've walked into your house. They give you as much room or as little room as you want, I think that's important.

David here describes a hospitable space, where comfort is achieved through small things and gestures; a network of non-human objects (a cup of tea, a biscuit), a semiotic rendering of domestic space and the human presences that animate that space. Maggie's commissions iconic pieces of architecture, and in these extra-ordinary buildings, everyday acts of kindness are routinely experienced and observed. These are acts of kindness that one will find in hospitals too (Riessman 2015), but the more hospitable spaces built by Maggie's can allow their users the room and time to reflect on their cancer in ways not easily available to them elsewhere (Butterfield and Martin 2016). Hospitals are not homogenous places and can be understood in a variety of ways by those who use them for different purposes (Street and Coleman 2012); nonetheless, they are spaces driven by clinical understandings of disease and increasingly shaped by market forces (Martin et al. 2015), whereas Maggie's Centres are underpinned by other understandings of cancer (for example, as a biographical point of departure) and thus offer something different in their environments. Through the materiality of its buildings, the work of staff and volunteers, and the experiences of visitors in need of support, Maggie's Centres evoke architectural atmospheres that *matter* to those who use them (Jacobs and Merriman 2011).

For an organisation celebrated for the spaces it provides, Maggie's gives their users the time to consider their changed circumstances in spaces which encourage a deeper reflection, alongside people that understand their condition, whether peers or professionals. In his critique of modern therapeutics drawn on above, Leder writes briefly about the idea of landscape within notions of disease:

> A landscape is viewed not as a field of possibility but of difficulties to be negotiated. The ordinary sense of free and spontaneous movement is now replaced by calculated effort; one does not want to take chances. Etymologically, 'ease' comes from the French word aise, originally

> meaning 'elbow room' or 'opportunity'. This experience of world-as-opportunity is precisely what dis-ease calls into question.
>
> (1990, p.81)

In part, at least, this may be where the value of Maggie's Centres resides for their visitors, in their *easing* open of opportunities to understand the emotional landscapes of cancer differently.

In Riessman's account of her cancer treatment she notes the importance of environments that afforded opportunities to move outside of her present situation and 'to take a momentary vacation from [her] identity as a cancer patient', in order to re-connect with her past identities (2015, p.1060). Moreover, working with occupational and physical therapists painfully through activities that initially prompted her resistance, for fear of their physical demands, meant that, in her words, they

> guided me towards an imagined future that they could see but I couldn't… Their minute actions inscribed a future onto my present, to which I – after much kicking and screaming – became committed. I could then become a partner in my own recovery.
>
> (2015, pp. 1064–1065)

Riessman's account is a hopeful one, pointing not merely to her physical recovery but also her ability to envisage a sense of the future, informed by her past identity and with a full recognition of her present situation. In articulating a sense of what geographies of hope might entail, Anderson suggests that we begin with 'a set of simple questions: what can a body do when it becomes hopeful? What capacities, and capabilities, are enabled?' (2006, p.734). Riessman's narrative account gives us a direct response to these questions. There is something resonant, too, in the temporalities through which Riessman reconstructs her biographical account, without closure and conditional as it remains, and the chronological dynamics that Anderson and Holden use to discuss how hope operates more widely, in the lives of cities (2008). In this work, they define hope as emerging 'in relation to the taking place of events' (p.144), and where the anticipation of future events can 'achieve effects over the here and now in relation to an event that has occurred' (p.155). Can these temporal dynamics of hope be translated from the scale of a city to the experience of the individual, and can the affects of extraordinary events be compared to the working out of everyday troubles, and their affects (Stewart 2007)?

In posing such questions I do not offer definitive answers, but there seems to me to be a particular importance in linking debates about the geographies of comfort alongside arguments about the atmospherics of hope, in order to open out the spatial potentialities that taking comfort seriously can offer (Bissell 2008). In reading through Riessman's narrative of her cancer diagnosis and treatment, it is striking how often the idea of comfort (alongside that of familiarity) recurs: to give just two examples, she writes of the pleasures of domestic life that revolved around 'the familiarity, the ease of habitual routines that comforts' (p.1061), and also the pain of leaving 'an interactional environment that had become familiar and comforting' (p.1061). This environment is the radiation unit, which she elsewhere characterises in terms of 'the cheerful greetings between patients and staff, the organisation of space in the waiting room, the talk between patients that produced an instant commonality, and the respect patients showed to those who didn't want to talk' (p.1060). Within this hospital setting, she sketches the atmospherics of care and hopeful occurrences observed time and again in Maggie's Centres. As noted above, in her hospital notes, Riessman found a means to inscribe 'a counter-narrative to the context-stripping medical record' (p.1062); relatedly, what we sense in Maggie's Centres are counter-geographies to the more typical environments in which visitors receive care. Rather, what we find are hopeful spaces in which they can take comfort.

Acknowledgements

This research was supported by funding from the Foundation for the Sociology of Health and Illness. Earlier drafts of the paper have been presented at the 2013 RGS-IBG Conference (London), a 2014 THESP seminar (Leeds) and the 2015 Materialities of Care Conference (York): the final draft has benefitted from comments at all events. I would like to thank the editors for their feedback and Sarah Nettleton for her valuable comments on the paper, as well as her support throughout this research. At Maggie's, I owe many thanks to Lesley Howells, the Research Lead, for her encouragement throughout this project, and also all to centre heads, staff members, volunteers and visitors who spoke to me about their experiences of using the Centres.

References

Anderson, B. (2006) Becoming and being hopeful: towards a theory of affect. *Environment and Planning D*, 24, 733–752.

Anderson, B. and Holden, A. (2008) Affective urbanism and the event of hope. *Space and Culture*, 11, 142–159.

Bille, M. (2015) Lighting up cosy atmosphere in Denmark. *Emotion, Space and Society*, 15, 56–63.

Bissell, D. (2008) Comfortable bodies: sedentary affects. *Environment and Planning A*, 40, 1697–1712.

Blaxter, M. (1978) Diagnosis as category and process: the case of alcoholism. *Social Science & Medicine*, 12, 9–17.

Blaxter, M. (2009) The case of the vanishing patient? Image and experience. *Sociology of Health & Illness*, 31(5), 762–778.

Brown, N. (2015) Metrics of hope: disciplining affect in oncology. *Health*, 19, 119–136.

Butterfield, A. and Martin, A. (2016) Affective sanctuaries: understanding Maggie's as therapeutic landscapes. *Landscape Research*, 41(6), 695–706.
Degen, M., DeSilvey, C. and Rose, G. (2008) Experiencing visualities in designed urban environments: learning from Milton Keynes. *Environment and Planning A*, 40, 1901–1920.
Frank, A. (1995) *The wounded storyteller: body, illness and ethics*. Chicago, IL: University of Chicago Press.
Jacobs, J. (2006) A geography of big things. *Cultural Geographies*, 13, 1–27.
Jacobs, J. and Merriman, P. (2011) Practising architectures. *Social & Cultural Geography*, 12(3), 211–222.
Jencks, C. (2010) The architecture of hope. In Jencks, C. and Heathcote, E. (eds.) *The architecture of hope*. London: Frances Lincoln, 8–43.
Jencks, C. and Heathcote, E. (2010) *The architecture of hope*. London: Frances Lincoln.
Jutel, A. (2009) Sociology of diagnosis: a preliminary review. *Sociology of Health & Illness*, 31(2), 278–299.
Jutel, A. and Nettleton, S. (2011) Towards a sociology of diagnosis: reflections and opportunities. *Social Science & Medicine*, 73, 793–800.
Keswick, M. (1995) Breast cancer: a view from the frontline. *The Breast*, 4(3), 205–210.
Kraftl, P. and Adey, P. (2008) Architecture/affect/inhabitation: geographies of being-in buildings. *Annals of the Association of American Geographers*, 98(1), 213–231.
Leder, D. (1990) *The absent body*. Chicago, IL: University of Chicago Press.
Leder, D. (1992) A tale of two bodies: the Cartesian corpse and the lived body. In Leder, D. (ed.) *The body in medical thought and practice*. Dordrecht: Kluwer Academic, 17–35.
Lees, L. (2001) Towards a critical geography of architecture: the case of an ersatz Colosseum. *Ecumene*, 8(1), 51–86.
Lees, L. and Baxter, R. (2011) A 'building event' of fear: thinking through the geography of architecture. *Social & Cultural Geography*, 12(2), 107–122.
Martin, D. (2016) Curating space, choreographing care: the efficacy of the everyday. In C. Bates, R. Imrie and K. Kullman (eds.) *Design as caring in an urban world*. Oxford: Wiley-Blackwell.
Martin, D. et al. (2015) Architecture and health care: a place for sociology. *Sociology of Health & Illness*, 37(7), 1007–1022.
Massey, A. (2011) *Chair*. London: Reaktion.
Mol, A. (2002) *The body multiple: ontology in medical practice*. Durham, NC: Duke University Press.
Mol, A. (2006) *The logic of care: health and the problem of patient choice*. London: Routledge.
Olson, R.E. (2011) Managing hope, denial or temporal anomie? Informal cancer carers' accounts of spouses' cancer diagnoses. *Social Science & Medicine*, 73, 904–911.
Pallasmaa, J. (2005) *The eyes of the skin: architecture and the senses*. Chichester: Wiley.
Paterson, M. (2011) More-than-visual approaches to architecture. Vision, touch, technique. *Social & Cultural Geography*, 12(3), 263–281.

Petersen, A. and Wilkinson, I. (2015) The sociology of hope in contexts of health, medicine and healthcare. *Health*, 19, 113–118.

Riessman, C.K. (2015) Ruptures and sutures: time, audience and identity in an illness narrative. *Sociology of Health & Illness*, 37(7), 1055–1071.

Rose, G., Degen, M. and Basdas, B. (2010) More on big things: building events and feelings. *Transactions of the Institute of British Geographers*, 35, 334–349.

Roth, J. (1963) *Timetables: structuring the passage of time in hospital treatment and other careers*. New York: Bobbs-Merrill.

Schaepe, K.S. (2011) Bad news and first impressions: patient and family caregiver accounts of learning the cancer diagnosis. *Social Science & Medicine*, 73, 912–921.

Schillmeier, M. and Domènech, M. (2009) Care and the art of dwelling: bodies, technologies, and home. *Space and Culture*, 12, 288–291.

Schillmeier, M. and Heinlein, M. (2009) Moving homes: from house to nursing home and the (un-)canniness of being at home. *Space and Culture*, 12, 218–231.

Stewart, K. (2007) *Ordinary affects*. Durham, NC: Duke University Press.

Street, A. and Coleman, S. (2012) Real and imagined spaces. *Space and Culture*, 15, 4–17.

Thrift, N. (2004) Intensities of feeling: towards a spatial politics of affect. *Geografiska Annaler Series B*, 86(1), 57–78.

Tuan, Y-F. (1974) *Topophilia: a study of environmental perception, attitudes, and values*. Eaglewood Cliffs, NJ: Prentice-Hall.

Willig, C. (2011) Cancer diagnosis as discursive capture: phenomenological repercussions of being positioned within dominant constructions of cancer. *Social Science & Medicine*, 73, 897–903.

Zerubavel, E. (1979) *Patterns of time in hospital life: a sociological perspective*. Chicago, IL: University of Chicago Press.

15 Feeling good, looking good

Comfort and the technologies of beauty in the spa

Jo Little and Katherine Morton

Introduction

The rapidly expanding literature on therapeutic geographies has seen a particular growth in work on healing and the places and practices of restoration and recovery (Williams, 2010; Foley, 2011). Within such work has been a recognition of the medicalisation or healthicisation of a range of physical and mental conditions (Lupton, 2003) relating to the everyday stress of living and to the increasing importance attached to general fitness and wellbeing. This has given rise, it is suggested, not only to new attitudes and treatments but to the development of an economy of healthcare spaces and consumer goods. With this paradigm shift in health has come a growing sense of personal responsibility for health and of the need to comply with a set of socially prescribed and accepted 'norms' in terms of the ways in which we care for our bodies and pay attention to our own wellbeing and for that of our families. We are encouraged, within this notion of personal responsibility, to ensure that we allow our bodies to rest and recuperate. Understanding how we make sense of and respond to these shifts in the meaning and treatment of health has been recognized in (post) medical geographies (Brown and Duncan, 2002) and beyond as highly relevant not only to the treatments we develop and apply but also to how these relate to broader ideas surrounding the care of the body and notions of identity and the self.

Developing from existing therapeutic geographies this chapter focuses on the practices of rest and relaxation and their role in individual fitness strategies and health routines. In so doing it explores the ways in which time to one's self and a routine of recuperation and relaxation practices have become essential to ideas of general wellbeing and to notions of recovery in terms of both particular illness and also the general pressures of daily life. The chapter looks specifically at the importance of rest and relaxation to women arguing that the medicalisation of particular conditions within the 'new paradigm of health', and the health practices employed in response to such conditions, are gendered (Moore, 2010). As such they reflect the continuation of important socio-cultural relations in the construction of masculinity and femininity and the performance of gender identities. This gendering is particularly apparent in the spaces of wellbeing and recovery that have developed to 'treat' issues of stress and the need for rest and relaxation, such as the spa and beauty salon. Indeed stress has emerged as a feminist issue, with implications for the gendered burdens of emotional labour, family health and domestic responsibilities.

In addition to examining the practices and spaces of rest and relaxation, the chapter will explore what we consider to be important (and again highly gendered) changes in the 'treatment' of tiredness. We argue that underlying the desire for, and engagement in, practices to aid rest and recovery are significant concerns about the appearance and fitness of the body and that such concerns are increasingly being reflected in the use of cosmetic procedures to treat the tired and ageing body. We suggest that pressures on women to maintain a particular corporeal size, shape and appearance as part of a normalised state of wellbeing is fuelling a major growth in technologies and spaces of cosmetic surgery and that the demand for time and pampering to assist the therapeutic practices of recovery are being increasingly replaced (or augmented) by specific and highly targeted cosmetic interventions. Such interventions treat the 'visibility' of fatigue and stress within the parameters of normative feminine corporeal ideals. Thus, in identifying this relationship between stress, tiredness and bodily appearance, we draw attention to the shifting boundaries between health and beauty and the blurring of the division between medical and cosmetic practices.

In exploring the relationship between medical and cosmetic practices in treating tiredness and promoting recovery and relaxation, our chapter focuses on the spa. As noted, the spaces where rest and recuperation take place have become key in the delivery of treatments. The huge growth in the market for and provision of spas is indicative of consumer desire for therapeutic responses to tiredness – responses that initially provided a space for escape and 'pampering' to attend to the tired body but which are increasingly designed to offer beauty treatments including cosmetic surgery to treat the visible signs of tiredness and stress. Drawing on original research on the use of the contemporary spa and on the cosmetic treatments now available at the spa we explore both the therapeutic practices themselves and also the motivations of women who engage in them. We look in particular at the importance attached to the embodied *appearance* of health and wellness and at how the management of a particular bodily size, shape, and appearance has become crucial to the regular health routines and broader experiences of wellbeing and comfort of many women. Before turning to the empirical part of the chapter and discussing the methodology used to collect the data, the chapter first develops some of the main theoretical directions shaping the study, including the gendered power relations surrounding ideas of health and fitness as well as the relationship between health and beauty in the understanding and treatment of ill health, stress and tiredness and the pursuit of comfort.

Stress and tiredness as issues of embodiment, wellbeing and identity

Geographies of health and wellbeing have long drawn attention to the healing properties of therapeutic landscapes that inspire quiet contemplation and reflection or simply convey peace (Kearns and Gesler, 1998). More recently the benefits of 'remote' and 'natural' landscapes have also included discussion of spiritual connections in the development of a broader perspective on the relationship between mental health and place. Conradson (2007: 33) for example, talks of the 'state of calm' and 'new forms of awareness' that may accompany human encounters with therapeutic landscapes. His work looks specifically at religious retreats and the 'experiential economy of stillness' associated with the therapeutic benefits of slowing down and creating time for mental reflection. As Conradson observes there is considerable scope for a more detailed examination of the issues associated with stillness and retreat in relation to health, in particular how the demand for restful and recuperative therapeutic practices relates to wellbeing, identity and everyday life.

The desire to seek out quiet spaces for relaxation as part of a strategy for healthy living is often seen as a response to the speed and business of daily life. It is argued that many people now face significantly more pressure at work than in the past and it is increasingly common for people to work very long hours. In addition, family life is frequently busy and complicated. Work, home and leisure take place at speed, reinforced by the presence of technology. While the pace of contemporary life provides much by way of stimulation, there is often little opportunity for 'time out' or even rest within busy and constantly changing schedules. Feminist geographers and sociologists have argued that women are disproportionally affected by the pressures of a busy life since they are still more likely to take primary responsibility for childcare in a family and are thus most likely to end up juggling employment and domestic duties (McDowell, 2005). Moreover, women's role as emotional carers means that they are more often burdened with the job of supporting all members of the family and broader friendship networks, and accommodating the pressures of such demands. The multi-layered nature of contemporary pressures (particularly on women) means that 'down time' cannot be accommodated within 'normal' everyday activities and environments but requires removal to places that are 'outside' (see Little, 2015).

The need for rest and recovery is clearly related not only to the growing place of 'stress' in contemporary society but also to the increasing understanding of the effects of stress on health. Jackson (2013, 16) provides a fascinating discussion of the historical development of stress as a 'hybrid phenomenon (that is), the product of both biological and cultural forces rendered visible by the technology and language of biomedical science'. He talks of the late twentieth and early twenty-first century as the 'age of stress' when people were increasingly recognised as subject to pressures at work and anxieties about the world around them. Documenting the rise of stress in terms of both social and biomedical theories Jackson explores the shifting terrain between stress and illness. He suggests that by the end of the twentieth-century stress had become an all-encompassing condition and a convenient vehicle for 'articulating and validating concerns about the attainability of happiness and wellbeing, about the impact of modern lifestyles on the balance of nature about the political security and economic stability of the world, and about the harmony of the cosmos' (Jackson, 2013, 20). There is not space here for a lengthy discussion of Jackson and other's accounts of contemporary understandings of stress. Rather, what we argue the most important here is the recognition of the ways in which the concept of stress powerfully associates health with the pursuit of relaxation and happiness. The stressed body is physically (and mentally) out of 'balance' and, as such, made vulnerable to ill health and hence unhappiness. What is also important here are the technologies and techniques that have been developed to mediate or eliminate the effects of stress – not only the medical treatment of stress but, increasingly, the alternative strategies that have been adopted as coping mechanisms.

Closely associated with stress, rest and recovery is the issue of sleep. Lack of sleep can be both a cause and a consequence of stress and, like stress, sleeplessness and tiredness have been absorbed into the medicalisation paradigm. According to Hislop and Arber (2003), sleep management has become problematised just as other 'normal' lifestyle factors such as diet and exercise and has also been 'discovered' as a potential risk factor in terms of future health. Very little has been written by geographers on the topic of sleep although, as Kraftl and Horton (2008) observe, there are a number of potentially interesting and fruitful aspects of sleep that should concern human geographers in relation particularly to issues of wellbeing, emotional health, embodiment and spatial and temporal rhythms and routines. It is also suggested that, in terms of emotional and embodied understandings of wellbeing, sleep – or at least the lack of sleep and the coping mechanisms used to offset the effects of tiredness – are gendered. Similar to the case of stress, women are more likely to suffer from sleep disruption and also to seek medical help to treat sleep 'problems'. Interestingly, Hislop and Arber (2003, 818) argue that in our relationship with sleep – in particular our entitlement to certain 'rights' such as freedom from noise and interference from others – we retain the status and authority (or lack thereof) associated with our daytime roles. In other words, sleep is deeply embedded in the social context of our lives such that 'women's attempts to meet the expectations of modern society in their working and domestic roles impose both restrictions on the time available for sleep as well as physical and psychological intrusions into their sleep.'

Despite the attention that has been paid to the healing properties of sleep, Williams (2005) argues that the medicalisation of sleep has been replaced by a 'healthicisation'. He suggests that sleep is increasingly being seen as a target for 'healthist cultures' with individuals acting as 'critical reflexive agents' in the management of sleep in the context of their health. There has been, so Hislop and Arber (2003) point out, evidence of a significant decline in the use of medication such as tranquillisers since the 1960s and 1970s with women, in particular, seeking out other ways to cope with sleep disruption. These authors talk about the increasing disillusion with the medical profession as a source of expertise in the treatment of sleep deprivation and the use of alternative self-directed strategies by those seeking to manage their sleep. As with the management of stress, a movement away from conventional forms of medical testament has seen an increase in practices of relaxation, such as meditation and yoga.

Central to this chapter is the association between stress and tiredness and the physical appearance of the body. Part of the interest in and concern for stress as a key aspect of wellbeing is its obvious manifestation in the way the body looks – often expressed in terms of its ageing effects. Bodies that appear tired are frequently framed as 'ill', 'unhealthy' or simply 'old', challenging acceptable norms of bodily appearance and self-care. A tired and/or ageing appearance is often associated with a body that is unruly or out of control. Again, the relationship between stress, tiredness, ageing and the healthy body, is in part gendered due to the strength of particular forms of the acceptable and desirable body in contemporary notions of femininity. The physical appearance of tiredness has, perhaps ironically, become of more concern as technologies have developed to modify and reduce the 'signs' of stress and ageing. While techniques of rest and relaxation (such as yoga) have become popular as ways of attending to stress and tiredness, a range of cosmetic (sometimes surgical) practices and treatments have also developed as a response to what are often seen as the more 'permanent' or 'visible', manifestations of the ageing body. Increasing pressure, it may be argued, is now placed on women, in particular, to avoid showing signs of tiredness and ageing through the normalisation of such cosmetic/surgical options although simultaneously, the ability to respond to pressures to look 'good' (i.e. 'young') may also be seen by some women as liberating and comforting; regaining 'control' over stress, tiredness and the ageing process.

The growing availability and use of cosmetic surgery to treat tiredness, stress and ageing relates to a fast-developing area of conceptual work on the body, health and identity. Such work has drawn attention to the growing acceptance of a range of cosmetic/beauty practices and their 'moral authorization' (Edmonds, 2013) as routine management of women's health. Indeed, as Edmonds and Sanabria (2014) observe in discussion of breast augmentation surgery in Brazil, certain treatments have become so much a part of everyday self-care that they are no longer considered to be medical enhancements. They identify the 'medical borderlands' where health and beauty treatments are entangled and where boundaries between health and beauty and between nature and technology are troubled. As many of these studies have noted, understanding the growing popularity and availability of cosmetic surgery practices involves a broader exploration of the ways in which they appear to respond to the 'pressures, anxieties and aspirations' of everyday life. Concerns raised about biopolitics and the relationship between technology and the body must also be seen in the light of ideas about the plasticity of the body and its ability to be changed and improved (Sanabria, 2011).

There is insufficient space here for the further development of these ideas or for a discussion of the many aspects of the body industry that have grown up around these 'borderlands' of medicine and beauty. What is important to emphasise for this chapter is the ways in which the cosmetic regimes of self-care seen as essential to modern femininity include not only practices of rest, relaxation and therapy but, increasingly, more invasive 'medical' treatments. The use of such treatments as acceptable and legitimate forms of wellbeing suggests the moral authorisation of aesthetic goals as health goals. Interestingly, as we discuss below, while beauty/aesthetic therapy has been seen as a *treat* to help women manage their busy lives, the stress associated with such lives is increasingly being seen as something that must be *treated* through medical/cosmetic procedures.

Self-care and the comfort of women's health and beauty regimes

The chapter now turns to consider these ideas in the light of empirical research. As explained in the brief discussion of methodology below, we draw on the findings of two distinct but related research projects to illustrate the importance of practices of rest and relaxation to women's health and wellbeing regimes. We also demonstrate how such regimes are increasingly/routinely making use of more invasive and technologically sophisticated cosmetic practices to respond to a need not only to *feel* good but also to *look* good. Through the presentation and discussion of our research we hope to unsettle the idea of 'comfort' in the context of women's health and wellbeing and show how it needs to be understood in terms of both the therapeutic practices themselves but also their wider relationship to issues of identity, femininity and the body.

Methodology

The research discussion starts by looking at women's wellbeing practices in the spa. Here we draw on a study of wellbeing and the contemporary spa undertaken in 2009/10. The original data were collected through a questionnaire of 50 spa users at a large hotel-based spa complex in Gloucestershire, UK. The questionnaire asked women about their use of the spa, the practices and treatments they had experienced and their motivations for going to the spa. This questionnaire was followed up with 5 in-depth interviews with spa managers and fitness workers (from the same spa and 2 others in Devon, UK) who discussed more general trends they had witnessed in spa use and popularity and also their strategies as therapeutic providers for responding to the shifting needs of their clients. All interviews were conducted at the spas and each lasted over 1.5 hours. The research discussion also makes use of ethnographic data based on participant observation was collected from 2 further spas.

The chapter also draws on data produced through a separate piece of research into the anti-ageing industry (Morton, 2014a) specifically focusing on the women's embodied engagement with anti-ageing technologies, the hybrid spaces of cosmetic and medical rationales and the professional identities and knowledges associated with such practices. The research took place between October 2011 and September 2012. The data is drawn from ethnographies of two aesthetic clinics, a beauty salon and a beauty training college in South-West England. This ethnographic data was also enriched through a series of semi-structured interviews. The interviews were vital for gaining a deeper insight into the experiences and knowledges of anti-ageing practices by both practitioners and consumers. In total 35 interviews were conducted, 16 with practitioners and 19 with consumers enrolled in the management and modification of ageing corporeality in various ways. The interviews lasted between one and two hours and took place in a variety of locations, including treatment rooms, back offices of treatment premises, homes and cafes.

While these were separate pieces of research they shared a common feminist ethical stance giving voice to issues surrounding women's bodies, emotions and subjectivities. Research was carried out in a sensitive and respectful way under the University of Exeter's research ethics guidelines. All participants were fully briefed about the purpose of both research studies and took part in the respective questionnaires and interviews knowingly and willingly.

Popularity of the spa (*'Because I Deserve It'*)

While relatively little in-depth academic research has been conducted on the use and popularity of the spa, the studies that have been done (mainly from business/marketing/tourism perspectives) (McNeil and Ragins, 2005) have drawn attention to the sharp rise in the numbers of spa goers over recent years. Spa visits, it is suggested, have become much more commonplace and regarded as a 'normal' part of leisure and tourist activity rather than the preserve of the rich and those recovering from illness or addiction. All the spa managers interviewed spoke of the popularity of their spa and the success of their business. Indeed two were actively pursuing expansion plans. The Gloucestershire spa offered a membership and also received day customers while the Devon spas operated no membership system. The Gloucestershire spa, at time of interview, had been open for five years and was full to capacity;

> We filled the membership before we even opened the door. We have 500 members with a further 570 in the waiting list. We are reasonably priced (membership is £105 a month and £85 for weekday with a joining fee of £150).
>
> (Glos. spa)

The questionnaire of spa goers that we draw on here demonstrated the extent to which respondents were committed to regular visits and while it is not surprising that the spa goers questioned were enthusiastic about the spa, the extent to which they were committed to very regular attendance was less anticipated and perhaps more noteworthy. Of the 50 respondents, 21 claimed to visit the spa at least once a week. The point here is not only the popularity of the spa as a form of leisure activity but, more pertinent to our arguments, how central spa visits appeared to have become in the health and beauty regimes of women questioned. Nearly half saw their spa visits as a 'high priority'. Moreover, 80 per cent of the spa users questioned claimed that they would come to the spa more often if they could. Constraints on their visits were mainly to do with time and money.

Further exploration of the free form comments of questionnaire respondents suggests that spa visits, for some, represent the kinds of spaces of relaxation and 'stillness' discussed by geographers as an important form of therapeutic landscape as noted above. For many women spa visits clearly bought them 'time out' and away from the family and from work and other responsibilities, providing a crucial opportunity for escape and 'me time'. When asked why she visited the spa one questionnaire respondent replied:

> to make me more relaxed and as a reward for working hard in my daily life.

While according to the spa managers:

> Our cliental is mainly working women who want to have a bit of fun and look after themselves. The escape element is very strong – even escape from humdrum lives as well as stressful lives.... This is their little oasis. Lots have the opportunity to join with partners or children but they chose not to because this is their space.
>
> (Glos. spa)

> We wish to provide an environment that allows you to escape from the tension and stress.
>
> (Devon spa)

Clearly, however, one of the chief functions of the spa in this context was to combine the opportunity for time to one's self with an element of indulgence. The spa thus combined time to focus on themselves with a sense of luxury and spoiling.

> We deliver a lovely product. There's no question, it does relax. From simple things like taking people out of their everyday lives and putting them in an environment where cups of tea are brought and someone is looking after you. We call it guilt-free relaxation.
>
> (Devon spa)

Many of the questionnaire respondents emphasized in their answers the extent to which their spa visit was a treat but one which they *deserved*. They noted the rejuvenating effects of the plush environment, the fluffy towels and robes and the quality of the cosmetics. In two of the spas visited, location added to that sense of luxury and reward with the spa setting providing an environment that was calming, beautiful and special. Typically spa visits have be seen as a reward or gift, marking a specific achievement and celebrating key life events. However, what this spa research demonstrated was the sense in which this notion of reward was being seen as an essential part of women's health and wellbeing regimes. Moreover, regular regeneration and recovery regimes *required* indulgence to be effective.

These findings support those of Black (2002) and Black and Sharma (2001) in their research on pampering and the body. Black (2002) claimed that the women she encountered in her study placed significant value on the therapeutic effects of pampering as a part of their routine body maintenance. She also argued that many women saw pampering as an important and increasingly regularized aspect of their health care and essential to their identities as professional women and to their ability to perform successfully in the workplace. As in the spas and beauty salons studied by Black (Ibid.), emphasis was placed in the spas studied here on the quality of the actual treatments offered. The spa managers talked at length about the choice of high quality products in the spa. Emphasis was placed on the choice of 'natural' products (possibly organic) that added a sense of luxury but that were affordable.

Coupled with the idea of escape and relaxation, the spa managers spoke of the spas as providing a nurturing environment. Two of the interviewees noted the benefits of the spa to people who were grieving or recovering from long term illness. Again, the effectiveness of the spa in meeting the needs of clients seeking this kind of recovery and nurturing was very much a function, according to the interviewees, of the quality of the spa setting and interior décor. The idea of repair being linked to luxury and pampering was strong and included, for both managers and consumers alike certain challenges in terms of the management of what were seen at times as competing demands. In particular, the challenge of accommodating the increasingly popular parties of one-off spa clients while maintaining the ambience and the atmosphere of relaxation and luxury for the individual and regular consumers was considered critical by the spa managers and was clearly a source of concern for the clients.

The importance of the spa as a place of sociability is interesting here beyond the challenges it posed for the management of different groups and the potential to change the spa experience. The interviewees all spoke at length about the spa as a place of bonding for friends and family groups – in particular mothers and daughters. The comfort provided by the spa was seen as important in the creation of spaces of celebration but also safe spaces of intimacy. One spa manager talked of the efforts she made at her spa to ensure people felt comfortable. As she noted:

> The worst experience you can have is going to the spa for the first time and being made to feel uncomfortable or unsure of what you should do. People need to know what they should wear and what they should take off or not. So people don't feel embarrassed.
>
> (Glos. spa)

The construction of the spa as a space of sociability also suggests a strong sense of shared or common goals. The popularity of spa visits as celebrations and as regular treats underpins the extent to which the spa responds to collectively held (and very powerful) norms about the appropriate body and the relationship between pampering, health and beauty. Again, the gendering of these norms is highly visible in the kinds of groups that use the spa (especially hen and bridal parties and women's friendship groups).

Beauty, body work and the spa

Significantly in relation to this chapter, the therapeutic value of spa visits was not simply articulated in terms of rest and recuperation but also involved perceived changes to appearance as a result of 'body work'. The contemporary spa has, for some time, clearly been more than the traditional baths or hot tub. Spas now offer a range of treatments and procedures aimed at 'enhancing', 'improving' and even 'correcting' the appearance of the body. Spa 'packages', popular as gifts, usually incorporate a choice of 'treatments' to accompany the time spent in saunas and hot tubs. For many of the women questioned in the research, these beauty treatments were a very important element of their spa visits with 54 per cent saying that they regularly used at least one of the treatments on offer in the Gloucestershire spa. One of the spa managers described how they were increasing their treatments by offering them later in the evening. As she explained – 'it's great for working women as it's the time you need to go for a treatment.'

Again, such treatments appeared to assume the role of both routine body maintenance and irregular treat. In so doing they also demonstrate the blurring of the boundaries between the medical and cosmetic, between health and beauty. Looking good (in terms of conforming to widely agreed norms of body size and appearance) was, for some women, inseparable from their sense of wellbeing and from the beneficial effects of relaxation and indulgence provided by the visit to the spa. Thus women arrived at the spa seeking rest and recovery yet strongly believed that they should leave looking 'better' than when they arrived. The comforting and relaxing effects of the spa would, it seems, have been diminished if clients had not been able to attend to what was seen as 'essential' body maintenance. As noted by Edmonds (2013) and referred to above, cosmetic treatments of all types have assumed a moral authority as essential to health and to the construction and performance of contemporary femininity.

One of the spa managers described how in the development of their spa facilities, additional treatment rooms had been fundamental. His spa offered 'spa breaks' where clients could come and stay in purpose-built accommodation and use the spa several times over the period of their stay (generally overnight but sometimes for up to a week). The spa manager noted that women visiting the spa would need to be able to have their 'regular' and 'normal' beauty treatments available as part of their spa therapies. This had meant providing additional 'services' such as manicures and leg waxing – initially as add-ons but increasingly as regular treatments available for all spa visitors.

A further critical element of the provision of beauty treatments as a regular part of the spa experience is the collective 'buy in' to the particular practices provided. Each of the spas studied as part of the research offered a basic set of beauty therapies which included facials, massages (often with hot stones), manicures and pedicures, eyelash tinting and hair removal. Added to this many also provided more specialist aromatherapy treatments. Such treatments were not an added extra to the regular spa visit but, as already noted, a very critical part.

For the Gloucestershire spa, the treatments represented the major part of their business (which also included a gym and fitness studios in addition to the spa, sauna and swimming pool). The manager spoke of the importance of responding to client's 'needs' in the planning and prioritizing of the spa's range of beauty treatments. Although a recently trialled Botox clinic at the spa had not received much support from regular spa members, the manager wanted to retain it 'in case people *needed* it' (emphasis added). She went on:

> If you are using this as your spa and fitness centre then you really ought to be having your regular treatments here. We have consciously done more treatments than some spas because if you are going to come here and have your massage and your facial we will also do your waxing and your nails.
>
> (Glos spa)

While the suggestion here is that the emphasis on beauty treatments is a practical issue it nevertheless makes a clear and powerful assumption that women will be undertaking these practices as part of their normal routines. There is no question that these are treatments that women will/should habitually undertake as part of their desire to conform to 'normal' gender identities and constructions of femininity.

The role of the beauty treatments as a key element of the spa experience and the importance of looking good as a part of feeling good were underlined by comments about the quality of beauty treatments on offer at the spas. Managers of both the Gloucestershire and the Devon spas were quick to distance themselves from the 'average' spa by asserting the quality of the treatments they offered.

> Yes we do have competition but we have a super product. A lot of girls set up on their own but it's difficult to survive. It's the difference between a Mercedes saloon and a little Fiesta. We provide a facial that's an hour and you won't get a better one anywhere in the world.

One manager recounted her experience of another spa:

> The children treated me to a day at (name of spa). I had a facial but I felt cheated because it was nothing I couldn't have done in front of the mirror... if you're going to pay for something and lie down for something then you want more than a half hour facial. So I don't do a half hour facial (at her spa) because I think people feel cheated, whether they are paying or not

The availability of 'experts' not only in terms of the beauty treatments but also for diet and nutrition, fitness, exercise was also seen as an important part of the spa experience for all the managers interviewed. Again this reinforces the sense in which the spa was seen not simply as a place of relaxation and escape, but as part of a broader package of self-care and bodily maintenance. Spa clients were encouraged not only to undertake beauty therapies and fitness classes but to develop personalized and holistic programmes of exercise, wellbeing and even life-coaching which incorporated relaxation and rest into the more specific diet, fitness, and cosmetic treatments.

In the next section of the chapter, we draw on the second piece of research in developing the links between health and beauty in relation to wellbeing therapies and practices. Our argument, as articulated at the outset of the chapter, is that the spa is increasingly shifting from a space devoted primarily to escape, comfort and relaxation to a space which includes a greater emphasis on cosmetic intervention. This shift we see as part of the normalization of certain beauty treatments and technologies and, as we shall show, a 'natural' extension of the role of (highly gendered) assumptions and expectations around the appearance of the body to feelings of health and wellbeing. Here, in line with the book's focus on comfort, we examine the visible and embodied effects of tiredness and on the cosmetic practices that seek to reduce the perceived appearance of tiredness and stress through beauty treatments.

Cosmetic intervention, visible ageing and self-transformation

Increasingly, the focus of efforts to treat stress and tiredness is turning to the appearance, or *visibility*, of such issues. In attending to the visibility of stress and tiredness, the cosmetic intervention industry offers a range of treatments. In the context of beauty salons, spas and aesthetic treatment clinics, the beauty industry reinforces gendered responsibilities to attend to the management of the appearance of tiredness, stress and more broadly ageing. Simultaneously the industry also offers 'solutions' to these issues, treatments which operate on both the long and short term visibility of signs of stress, fatigue and ageing. The focus of this industry is increasingly on 'repair', 'rejuvenation' and 'refreshment' of skin bearing the marks of busy lifestyles, 'unhealthy' habits, and lack of sleep. Complexly interwoven with the functional, aesthetic-driven goals of such treatments, they are also framed in similar ways to the beauty or pampering treatments practised in the spa; part of the experience of such treatments is framed in terms of taking time out, relaxation, 'me time' and treating the self. Such treatments, and the spaces through which they are enacted, relate productively to the medicalisation of stress and the emphasis on certain practices. In similar ways to the spa, we can see here efforts to tackle, treat, maintain and reshape the 'excessive', 'unruly' feminine body to conform to particular feminine corporeal norms and expectations around control.

Within spaces of cosmetic intervention, the body-work technologies and practices associated with 'treating' the visible signs of ageing, stress and fatigue vary, for instance in terms of the extent of 'invasiveness', technology type and cost. 'Treatments' include anti-ageing facials, electrical facials, cosmeceuticals and non-surgical procedures such as laser rejuvenation, chemical peels and injectables such as Botox®. Analysis of interview data suggested that consumers, primarily women, seek these treatments for a variety of reasons, from insecurities and anxieties about a particular aspect of their appearance, in responses to a stressful or particularly emotional period of their lives, a desire to 'renew' themselves following illness, a break-up or a 'significant' birthday, to name a few (Morton, 2014b). Such rationales were further imbued with emotional connotations, with consumers reflecting on the perceived 'disjuncture' between their appearance and how they felt;

> These lines [gestures to around eyes] make me look tired and quite sad I think. They give off the impression that I am worn out, that I feel worn out because my face looks worn out.
>
> (Consumer Interview Twelve: Joss, 28, Legal Secretary)

> I look angry all the time, cross and frowning, and people don't respond to that well. I don't want to look tired or sad or cross anymore…that's not me
>
> (Consumer Interview Seven: Liz, 61, Foster Carer)

Such discussions were prevalent throughout the research, indicating the complexity of visible signs of ageing, stress, and fatigue in relation to self-perception and identity, conflating the self and appearance. In similar ways to the spa, therefore, the consumption of techniques and practices within spaces of cosmetic intervention was framed not only in terms of treating the pathological ageing or fatigued appearance (Straughan, 2010), but simultaneously in terms of transforming the self, through 'me-time' and relaxation, in a treatment space framed as distinct from the pressures and demands of everyday life (Black, 2002; Little, 2013);

> The treatment (anti-ageing facial) is really special, its luxurious and the products we use are adapted from dermatological research into skin cancer so its real medical-grade stuff...the results are instant and get better the more frequently clients come and see us, and with us they get that really relaxing treatment then they leave back to the real world at peace.
>
> (Practitioner Interview Seven: Mel, Beauty Therapist)

Interestingly, here, and in contrast to the spa, 'me-time' often included treatments which involved discomfort and even pain, which were carried out in the same vein as a 'treat' or a 'luxury' experience at the spa, enabling women to feel 'better in themselves through the promotion of 'aesthetic capital' (Anderson et al. 2010, 565) and attending to the 'improvement' of their appearance and fulfilling the moral responsibilities associated with health, wellbeing and self-care (Moore, 2010). Such treatments were also framed within the interview data in terms of facilitating a re-alignment of the individual's appearance and sense of self, remedying the discomfort experienced for instance by the appearance of ageing. Thus cosmetic intervention is not only an aesthetic practice, but a practice perceived by consumers as having the capacity to transform them back to the person they really are, framed in Foucauldian terms as 'self-actualisation' (Heyes, 2007; Gimlin, 2012). Such practices have become increasingly normalized, regular and expected, whilst simultaneously framed as a relaxing treat.

Conclusions

Both the spa-based research and this attention to cosmetic intervention have focused on the increasing ubiquity of cosmetic treatments and technologies within narratives of self-care, health and wellbeing. Such attention demonstrates the conflation of issues of tiredness, stress and ageing, and the ways in which the 'appropriate' or 'acceptable' female body is recognized as synonymous with the 'youthful' body. Both aspects of the research highlight the significance of the *visibility* of fatigue, stress and ageing to individuals. The research also highlights the ways in which spaces of body-work are becoming more medicalised, employing more sophisticated technologies in treating the pathologised appearance. Part of the experience of such spaces is about relaxation, but becoming more prominent are the functional/aesthetic outcomes of such treatments, with focus moving beyond 'rejuvenation' to 'enhancement' and as such the redefinition of what it means to age.

Thus we can observe an enfolding of the emotional, comforting dimensions of body-work in the spa, beauty salon and aesthetic clinic, with the reassuring, comforting dimensions of engagement with aesthetic practices to improve and normalize the feminine body and self-perception.

This chapter has drawn attention to the practices of pampering and relaxation in the spa and beauty salon. It has highlighted women's engagement with these practices as a means to reduce the physical and emotional effects of stress and tiredness. In doing so the chapter has demonstrated the ways through which the stressed, tired, and ageing body is managed and maintained, reinforcing the emotional significance of such body-work. This work contributes to ongoing interdisciplinary discussions about relaxation and the management of stress, and the increasing significance such issues have in relation to wellbeing. Furthermore, this chapter has provided analysis of a context in which socio-cultural norms and expectations around idealized feminine appearance are incorporated into gendered responsibilities around health and the maintenance of 'acceptable' corporeality, in the control and management of unruly bodies and emotions. Within a broader context of increasing technological capacitates to modify and mediate the body, this chapter has articulated the complex entanglements of appearance-work and the management of the physical and emotional dimensions of stress.

References

Anderson, T.L., Grunert, C., Katz, A. and Lovascio, S. (2010) 'Aesthetic Capital: A Research Review on Beauty Perks and Penalties', *Sociology Compass*, 4(8), 564–575.

Black, P. (2002) '"Ordinary People Come Through Here": Locating the Beauty Salon in Women's Lives', *Feminist Review*, 71, 2–17.

Black, P. and Sharma, U. (2001) 'Men are Real, Women Are "made up": Beauty Therapy and the Construction of Femininity', *The Sociological Review*, 49(1), 100–116.

Brown, T. and Duncan, C. (2002) 'Placing Geographies of Public Health', *Area*, 34(4), 361–96.

Conradson, D. (2007) 'The Experiential Economy of Stillness, Places of Retreat in Contemporary Britain', in Williams, A. (Ed.) *Therapeutic Landscapes*. Farnham: Ashgate, pp. 33–48.

Edmonds, A. (2013) 'The Biological Subject of Aesthetic Medicine', *Feminist Theory*, 14(1), 65–82.

Edmonds, A. and Sanabria, E. (2014) 'Medical Borderlands: Engineering the Body with Plastic Surgery and Hormonal Therapies in Brazil', *Anthropology and Medicine*, 21(2), 202–216.

Foley, R. (2011) 'Performing Health in Place: The Holy Well as a Therapeutic Assemblage', *Health and Place*, 17(2), 470–479.

Gimlin, D. (2012) *Cosmetic Surgery Narratives: A Cross-Cultural Analysis of Women's Accounts*. Basingstoke: Palgrave Macmillan.

Heyes, C.J. (2007) *Self-Transformations: Foucault, Ethics, and Normalised Bodies*. Oxford: Oxford University Press.

Hislop, J. and Arber, S. (2003) 'Understanding Women's Sleep Management: Beyond Medicalization-healthicization?', *Sociology of Health and Illness*, 25(7), 815–837.

Jackson, M. (2013) *The Age of Stress: Science and the Search for Stability*. Oxford: Oxford University Press.

Kearns, R.A and Gesler, W.M. (1998) *Putting Health into Place: Landscape, Identity, and Well-being*. Syracuse: Syracuse University Press.

Kraftl, P. and Horton, J. (2008) 'Spaces of Every-night Life: For Geographies of Sleep, Sleeping and Sleepiness', *Progress in Human Geography*, 32(4), 509–524.

Little, J. (2013) 'Pampering, Well-being and Women's Bodies in the Therapeutic Spaces of the Spa', *Social and Cultural Geography*, 14(1), 41–58.

Little, J. (2015) 'Nature, Well-being and the Transformational Self', *The Geographical Journal*, 181(2), 121–128.

Lupton, D. (2003) *Medicine as Culture*. London: Sage.

McDowell, L. (2005) 'Love, Money, and Gender Divisions of Labour: Some Critical Reflection on Welfare-to-work Policies in the UK', *Journal of Economic Geography*, 5(3), 365–379.

Moore, S. (2010) 'Is the Healthy Body Gendered? Toward a Feminist Critique of the New Paradigm of Health', *Body and Society*, 16(2), 95–118.

Morton, K. (2014a) *Anti-Ageing and Women's Bodies: Spaces, Practices, and Knowledges of Cosmetic Intervention*, Ph.D Thesis, University of Exeter. Available from: http://hdl.handle.net/10871/16000

Morton, K. (2014b) 'Emerging Geographies of Disciplining the Ageing Body: Practicing Cosmetic Technologies in the Aesthetic Clinic', *Gender, Place and Culture: A Journal of Feminist Geography*, DOI: 10.1080/0966369X.2014.939151

McNeil, K. and Ragins, E. (2005) 'Staying in the Spa Marketing Game: Trends, Challenges, Strategies and Techniques', *Journal of Vacation Marketing*, 11, 31–39.

Sanabria (2011) 'The Body Inside Out', *Social Analysis: International Journal of Social and Cultural Practices*, 55(1), 94–112.

Straughan, E. (2010) 'The Salon as Clinic: Problematizing, Treating, and Caring for Skin', *Social and Cultural Geography*, 11(7), 647–661.

Williams, A.M. (2010) 'Spiritual Therapeutic Landscapes and Healing: A Case Study of St. Anne de Beaupre, Quebec, Canada', *Social Science & Medicine*, 70(10), 1633–1640.

Williams, S (2005) *Sleep and Society: Sociological Ventures into the Unknown*. London: Routledge.

Index

Note: *Italic* page numbers refer to figures and page numbers followed by "n" denote endnotes.

www.ingramcontent.com/pod-product-compliance
Ingram Content Group UK Ltd.
Pitfield, Milton Keynes, MK11 3LW, UK
UKHW020151200926
475702UK00016B/2661

9780367682613